MW00511448

SCIENCE 400

Teacher's Guide

Author:
Alpha Omega Publications

Editor:
Alan Christopherson, M.S.

804 N. 2nd Ave. E.,
Rock Rapids, IA 51246-1759

SCIENCE 400

LIFEPAC® Overview

SCIENCE SCOPE & SEQUENCE

	Grade 1	Grade 2	Grade 3
UNIT 1	YOU LEARN WITH YOUR EYES • Name and group some colors • Name and group some shapes • Name and group some sizes • Help from what you see	THE LIVING AND NONLIVING • What God created • Rock and seed experiment • God-made objects • Man-made objects	YOU GROW AND CHANGE • Air we breathe • Food for the body • Exercise and rest • You are different
UNIT 2	YOU LEARN WITH YOUR EARS • Sounds of nature and people • How sound moves • Sound with your voice • You make music	PLANTS • How are plants alike • Habitats of plants • Growth of plants • What plants need	PLANTS • Plant parts • Plant growth • Seeds and bulbs • Stems and roots
UNIT 3	MORE ABOUT YOUR SENSES • Sense of smell • Sense of taste • Sense of touch • Learning with my senses	ANIMALS • How are animals alike • How are animals different • What animals need • Noah and the ark	ANIMAL AND ENVIRONMENT CHANGES • What changes an environment • How animals are different • How animals grow • How animals change
UNIT 4	ANIMALS • What animals eat • Animals for food • Animals for work • Pets to care for	YOU • How are people alike • How are you different • Your family • Your health	YOU ARE WHAT YOU EAT • Food helps your body • Junk foods • Food groups • Good health habits
UNIT 5	PLANTS • Big and small plants • Special plants • Plants for food • House plants	PET AND PLANT CARE • Learning about pets • Caring for pets • Learning about plants • Caring for plants	PROPERTIES OF MATTER • Robert Boyle • States of matter • Physical changes • Chemical changes
UNIT 6	GROWING UP HEALTHY • How plants and animals grow • How your body grows • Eating and sleeping • Exercising	YOUR FIVE SENSES • Your eye • You can smell and hear • Your taste • You can feel	SOUNDS AND YOU • Making sounds • Different sounds • How sounds move • How sounds are heard
UNIT 7	GOD'S BEAUTIFUL WORLD • Types of land • Water places • The weather • Seasons	PHYSICAL PROPERTIES • Colors • Shapes • Sizes • How things feel	TIMES AND SEASONS • The earth rotates • The earth revolves • Time changes • Seasons change
UNIT 8	ALL ABOUT ENERGY • God gives energy • We use energy • Ways to make energy • Ways to save energy	OUR NEIGHBORHOOD • Things not living • Things living • Harm to our world • Caring for our world	ROCKS AND THEIR CHANGES • Forming rocks • Changing rocks • Rocks for buildings • Rock collecting
UNIT 9	MACHINES AROUND YOU • Simple levers • Simple wheels • Inclined planes • Using machines	CHANGES IN OUR WORLD • Seasons • Change in plants • God's love never changes • God's Word never changes	HEAT ENERGY • Sources of heat • Heat energy • Moving heat • Benefits and problems of heat
UNIT 10	WONDERFUL WORLD OF SCIENCE • Using your senses • Using your mind • You love yourself • You love the world	LOOKING AT OUR WORLD • Living things • Nonliving things • Caring for our world • Caring for ourselves	PHYSICAL CHANGES • Change in man • Change in plants • Matter and time • Sound and energy

SCIENCE SCOPE & SEQUENCE

Grade 4	Grade 5	Grade 6	
PLANTS • Plants and living things • Using plants • Parts of plants • The function of plants	CELLS • Cell composition • Plant and animal cells • Life of cells • Growth of cells	PLANT SYSTEMS • Parts of a plant • Systems of photosynthesis • Transport systems • Regulatory systems	UNIT 1
ANIMALS • Animal structures • Animal behavior • Animal instincts • Man protects animals	PLANTS: LIFE CYCLES • Seed producing plants • Spore producing plants • One-celled plants • Classifying plants	BODY SYSTEMS • Digestive system • Excretory system • Skeletal system • Health and diseases	UNIT 2
MAN'S ENVIRONMENT • Resources • Balance in nature • Communities • Conservation and preservation	ANIMALS: LIFE CYCLES • Invertebrates • Vertebrates • Classifying animals • Relating function and structure	PLANT AND ANIMAL BEHAVIOR • Animal behavior • Plant behavior • Plant-animal interaction • Cycles and balance in nature	UNIT 3
MACHINES • Work and energy • Simple machines • Simple machines together • Complex machines	BALANCE IN NATURE • Needs of life • Dependence on others • Prairie life • Stewardship of nature	MOLECULAR GENETICS • Reproduction • Inheritance • DNA and mutations • Mendel's work	UNIT 4
ELECTRICITY AND MAGNETISM • Electric current • Electric circuits • Magnetic materials • Electricity and magnets	TRANSFORMATION OF ENERGY • Work and energy • Heat energy • Chemical energy • Energy sources	CHEMICAL STRUCTURE • Nature of matter • Periodic Table • Diagrams of atoms • Chemical changes	UNIT 5
PROPERTIES OF MATTER • Properties of water • Properties of matter • Molecules and atoms • Elements	RECORDS IN ROCK: THE FLOOD • The Biblical account • Before the flood • The flood • After the flood	LIGHT AND SOUND • Sound waves • Light waves • The visible spectrum • Colors	UNIT 6
WEATHER • Causes of weather • Forces of weather • Observing weather • Weather instruments	RECORDS IN ROCK: FOSSILS • Fossil types • Fossil location • Identifying fossils • Reading fossils	MOTION AND ITS MEASUREMENT • Definition of work • Rate or force of doing work • Laws of motion and gravitation • Change in motion	UNIT 7
THE SOLAR SYSTEM • Our solar system • The big universe • Sun and planets • Stars and space	RECORDS IN ROCK: GEOLOGY • Features of the earth • Rock of the earth • Forces of the earth • Changes in the earth	SPACESHIP EARTH • Shape of the earth • Rotation and revolution • Eclipses • The solar system	UNIT 8
THE PLANET EARTH • The atmosphere • The hydrosphere • The lithosphere • Rotation and revolution	CYCLES IN NATURE • Properties of matter • Changes in matter • Natural cycles • God's order	ASTRONOMY AND THE STARS • History of astronomy • Investigating stars • Major stars • Constellations	UNIT 9
GOD'S CREATION • Earth and solar system • Matter and weather • Using nature • Conservation	LOOK AHEAD • Plant and animal life • Balance in nature • Biblical records • Records of rock	THE EARTH AND THE UNIVERSE • Plant systems • Animal systems • Physics and chemistry • The earth and stars	UNIT 10

SCIENCE SCOPE & SEQUENCE

	Grade 7	Grade 8	Grade 9
UNIT 1	WHAT IS SCIENCE • Tools of a scientist • Methods of a scientist • Work of a scientist • Careers in science	SCIENCE AND SOCIETY • Definition of science • History of science • Science today • Science tomorrow	OUR ATOMIC WORLD • Structure of matter • Radioactivity • Atomic nuclei • Nuclear energy
UNIT 2	PERCEIVING THINGS • History of the metric system • Metric units • Advantages of the metric system • Graphing data	STRUCTURE OF MATTER I • Properties of matter • Chemical properties of matter • Atoms and molecules • Elements, compounds, & mixtures	VOLUME, MASS, AND DENSITY • Measure of matter • Volume • Mass • Density
UNIT 3	EARTH IN SPACE I • Ancient stargazing • Geocentric Theory • Copernicus • Tools of astronomy	STRUCTURE OF MATTER II • Changes in matter • Acids • Bases • Salts	PHYSICAL GEOLOGY • Earth structures • Weathering and erosion • Sedimentation • Earth movements
UNIT 4	EARTH IN SPACE II • Solar energy • Planets of the sun • The moon • Eclipses	HEALTH AND NUTRITION • Foods and digestion • Diet • Nutritional diseases • Hygiene	HISTORICAL GEOLOGY • Sedimentary rock • Fossils • Crustal changes • Measuring time
UNIT 5	THE ATMOSPHERE • Layers of the atmosphere • Solar effects • Natural cycles • Protecting the atmosphere	ENERGY I • Kinetic and potential energy • Other forms of energy • Energy conversions • Entropy	BODY HEALTH I • Microorganisms • Bacterial infections • Viral infections • Other infections
UNIT 6	WEATHER • Elements of weather • Air masses and clouds • Fronts and storms • Weather forecasting	ENERGY II • Magnetism • Current and static electricity • Using electricity • Energy sources	BODY HEALTH II • Body defense mechanisms • Treating disease • Preventing disease • Community health
UNIT 7	CLIMATE • Climate and weather • Worldwide climate • Regional climate • Local climate	MACHINES I • Measuring distance • Force • Laws of Newton • Work	ASTRONOMY • Extent of the universe • Constellations • Telescopes • Space explorations
UNIT 8	HUMAN ANATOMY I • Cell structure and function • Skeletal and muscle systems • Skin • Nervous system	MACHINES II • Friction • Levers • Wheels and axles • Inclined planes	OCEANOGRAPHY • History of oceanography • Research techniques • Geology of the ocean • Properties of the ocean
UNIT 9	HUMAN ANATOMY II • Respiratory system • Circulatory system • Digestive system • Endocrine system	BALANCE IN NATURE • Photosynthesis • Food • Natural cycles • Balance in nature	SCIENCE AND TOMORROW • The land • Waste and ecology • Industry and energy • New frontiers
UNIT 10	CAREERS IN SCIENCE • Scientists at work • Astronomy • Meteorology • Medicine	SCIENCE AND TECHNOLOGY • Basic science • Physical science • Life science • Vocations in science	SCIENTIFIC APPLICATIONS • Measurement • Practical health • Geology and astronomy • Solving problems

SCIENCE SCOPE & SEQUENCE

	Grade 10	Grade 11	Grade 12
UNIT 1	TAXONOMY • History of taxonomy • Binomial nomenclature • Classification • Taxonomy	INTRODUCTION TO CHEMISTRY • Metric units and instrumentation • Observation and hypothesizing • Scientific notation • Careers in chemistry	KINEMATICS • Scalars and vectors • Length measurement • Acceleration • Fields and models
UNIT 2	BASIS OF LIFE • Elements and molecules • Properties of compounds • Chemical reactions • Organic compounds	BASIC CHEMICAL UNITS • Alchemy • Elements • Compounds • Mixtures	DYNAMICS • Newton's Laws of Motion • Gravity • Circular motion • Kepler's Laws of Motion
UNIT 3	MICROBIOLOGY • The microscope • Protozoan • Algae • Microorganisms	GASES AND MOLES • Kinetic theory • Gas laws • Combined gas law • Moles	WORK AND ENERGY • Mechanical energy • Conservation of energy • Power and efficiency • Heat energy
UNIT 4	CELLS • Cell theories • Examination of the cell • Cell design • Cells in organisms	ATOMIC MODELS • Historical models • Modern atomic structure • Periodic Law • Nuclear reactions	WAVES • Energy transfers • Reflection and refraction of waves • Diffraction and interference • Sound waves
UNIT 5	PLANTS: GREEN FACTORIES • The plant cell • Anatomy of the plant • Growth and function of plants • Plants and people	CHEMICAL FORMULAS • Ionic charges • Electronegativity • Chemical bonds • Molecular shape	LIGHT • Speed of light • Mirrors • Lenses • Models of light
UNIT 6	HUMAN ANATOMY AND PHYSIOLOGY • Digestive and excretory systems • Respiratory and circulatory systems • Skeletal and muscular systems • Body control systems	CHEMICAL REACTIONS • Detecting reactions • Energy changes • Reaction rates • Equilibriums	STATIC ELECTRICITY • Nature of charges • Transfer of charges • Electric fields • Electric potential
UNIT 7	INHERITANCE • Gregor Mendel's experiments • Chromosomes and heredity • Molecular genetics • Human genetics	EQUILIBRIUM SYSTEMS • Solutions • Solubility equilibriums • Acid-base equilibriums • Redox equilibriums	CURRENT ELECTRICITY • Electromotive force • Electron flow • Resistance • Circuits
UNIT 8	CELL DIVISION & REPRODUCTION • Mitosis and meiosis • Asexual reproduction • Sexual reproduction • Plant reproduction	HYDROCARBONS • Organic compounds • Carbon atoms • Carbon bonds • Saturated and unsaturated	MAGNETISM • Fields • Forces • Electromagnetism • Electron beams
UNIT 9	ECOLOGY & ENERGY • Ecosystems • Communities and habitats • Pollution • Energy	CARBON CHEMISTRY • Saturated and unsaturated • Reaction types • Oxygen groups • Nitrogen groups	ATOMIC AND NUCLEAR PHYSICS • Electromagnetic radiation • Quantum theory • Nuclear theory • Nuclear reaction
UNIT 10	APPLICATIONS OF BIOLOGY • Principles of experimentation • Principles of reproduction • Principles of life • Principles of ecology	ATOMS TO HYDROCARBONS • Atoms and molecules • Chemical bonding • Chemical systems • Organic chemistry	KINEMATICS TO NUCLEAR PHYSICS • Mechanics • Wave motion • Electricity • Modern physics

STRUCTURE OF THE LIFEPAC CURRICULUM

The LIFEPAC curriculum is conveniently structured to provide one teacher handbook containing teacher support material with answer keys and ten student worktexts for each subject at grade levels two through twelve. The worktext format of the LIFEPACs allows the student to read the textual information and complete workbook activities all in the same booklet. The easy to follow LIFEPAC numbering system lists the grade as the first number(s) and the last two digits as the number of the series. For example, the Language Arts LIFEPAC at the 6th grade level, 5th book in the series would be LAN0605.

Each LIFEPAC is divided into 3 to 5 sections and begins with an introduction or overview of the booklet as well as a series of specific learning objectives to give a purpose to the study of the LIFEPAC. The introduction and objectives are followed by a vocabulary section which may be found at the beginning of each section at the lower levels, or in the glossary at the high school level. Vocabulary words are used to develop word recognition and should not be confused with the spelling words introduced later in the LIFEPAC. The student should learn all vocabulary words before working the LIFEPAC sections to improve comprehension, retention, and reading skills.

Each activity or written assignment has a number for easy identification, such as 1.1. The first number corresponds to the LIFEPAC section and the number to the right of the decimal is the number of the activity.

Teacher checkpoints, which are essential to maintain quality learning, are found at various locations throughout the LIFEPAC. The teacher should check 1) neatness of work and penmanship, 2) quality of understanding (tested with a short oral quiz), 3) thoroughness of answers (complete sentences and paragraphs, correct spelling, etc.), 4) completion of activities (no blank spaces), and 5) accuracy of answers as compared to the answer key (all answers correct).

The self test questions are also number coded for easy reference. For example, 2.015 means that this is the 15th question in the self test of Section 2. The first number corresponds to the LIFEPAC section, the zero indicates that it is a self test question, and the number to the right of the zero is the question number.

The LIFEPAC test is packaged at the centerfold of each LIFEPAC. It should be removed and put aside before giving the booklet to the student for study.

Answer and test keys have the same numbering system as the LIFEPACs and appear throughout this handbook. The student may be given access to the answer keys (not the test keys) under teacher supervision so that he can score his own work.

A thorough study of the Curriculum Overview by the teacher before instruction begins is essential to the success of the student. The teacher should become familiar with expected skill mastery and understand how these grade-level skills fit into the overall skill development of the curriculum. The teacher should also preview the objectives that appear at the beginning of each LIFEPAC for additional preparation and planning.

TEST SCORING AND GRADING

Answer keys and test keys give examples of correct answers. They convey the idea, but the student may use many ways to express a correct answer. The teacher should check for the essence of the answer, not for the exact wording. Many questions are high level and require thinking and creativity on the part of the student. Each answer should be scored based on whether or not the main idea written by the student matches the model example. "Any Order" or "Either Order" in a key indicates that no particular order is necessary to be correct.

Most self tests and LIFEPAC tests at the lower elementary levels are scored at 1 point per answer; however, the upper levels may have a point system awarding 2 to 5 points for various answers or questions. Further, the total test points will vary; they may not always equal 100 points. They may be 78, 85, 100, 105, etc.

Example 1

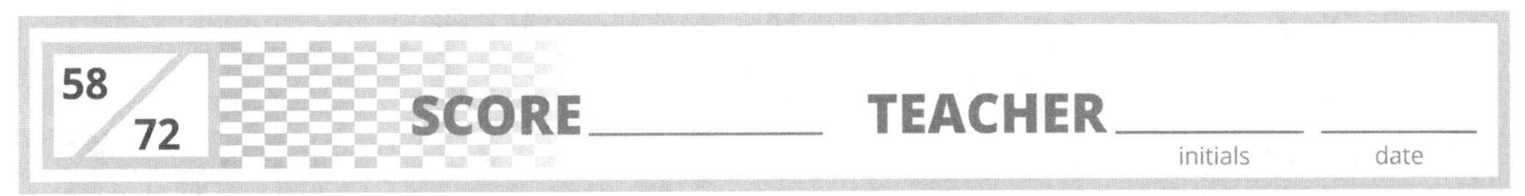

Example 2

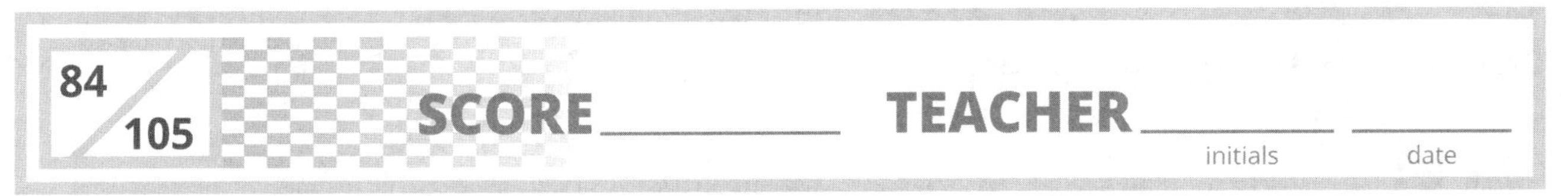

A score box similar to ex.1 above is located at the end of each self test and on the front of the LIFEPAC test. The bottom score, 72, represents the total number of points possible on the test. The upper score, 58, represents the number of points your student will need to receive an 80% or passing grade. If you wish to establish the exact percentage that your student has achieved, find the total points of his correct answers and divide it by the bottom number (in this case 72) For example, if your student has a point total of 65, divide 65 by 72 for a grade of 90%. Referring to ex. 2, on a test with a total of 105 possible points, the student would have to receive a minimum of 84 correct points for an 80% or passing grade. If your student has received 93 points, simply divide the 93 by 105 for a percentage grade of 89%. Students who receive a score below 80% should review the LIFEPAC and retest using the appropriate Alternate Test found in the Teacher's Guide.

The following is a guideline to assign letter grades for completed LIFEPACs based on a maximum total score of 100 points.

Example:

LIFEPAC Test	=	60% of the Total Score (or percent grade)
Self Test	=	25% of the Total Score (average percent of self tests)
Reports	=	10% or 10* points per LIFEPAC
Oral Work	=	5% or 5* points per LIFEPAC

*Determined by the teacher's subjective evaluation of the student's daily work.

Example:

LIFEPAC Test Score	=	92%	92 x .60	= 55 points
Self Test Average	=	90%	90 x .25	= 23 points
Reports				= 8 points
Oral Work				= 4 points
TOTAL POINTS				= 90 points

Grade Scale based on point system:

100 – 94	=	A
93 – 86	=	B
85 – 77	=	C
76 – 70	=	D
Below 70	=	F

TEACHER HINTS AND STUDYING TECHNIQUES

LIFEPAC activities are written to check the level of understanding of the preceding text. The student may look back to the text as necessary to complete these activities; however, a student should never attempt to do the activities without reading (studying) the text first. Self tests and LIFEPAC tests are never open book tests.

Language arts activities (skill integration) often appear within other subject curriculum. The purpose is to give the student an opportunity to test his skill mastery outside of the context in which it was presented.

Writing complete answers (paragraphs) to some questions is an integral part of the LIFEPAC curriculum in all subjects. This builds communication and organization skills, increases understanding and retention of ideas, and helps enforce good penmanship. Complete sentences should be encouraged for this type of activity. Obviously, single words or phrases do not meet the intent of the activity, since multiple lines are given for the response.

Review is essential to student success. Time invested in review where review is suggested will be time saved in correcting errors later. Self tests, unlike the section activities, are closed book. This procedure helps to identify weaknesses before they become too great to overcome. Certain objectives from self tests are cumulative and test previous sections; therefore, good preparation for a self test must include all material studied up to that testing point.

The following procedure checklist has been found to be successful in developing good study habits in the LIFEPAC curriculum.

1. Read the introduction and Table of Contents.
2. Read the objectives.
3. Recite and study the entire vocabulary (glossary) list.
4. Study each section as follows:
   a. Read the introduction and study the section objectives.
   b. Read all the text for the entire section, but answer none of the activities.
   c. Return to the beginning of the section and memorize each vocabulary word and definition.
   d. Reread the section, complete the activities, check the answers with the answer key, correct all errors, and have the teacher check.
   e. Read the self test but do not answer the questions.
   f. Go to the beginning of the first section and reread the text and answers to the activities up to the self test you have not yet done.
   g. Answer the questions to the self test without looking back.
   h. Have the self test checked by the teacher.
   i. Correct the self test and have the teacher check the corrections.
   j. Repeat steps a–i for each section.
5. Use the SQ3R method to prepare for the LIFEPAC test.

   Scan the whole LIFEPAC.
   Question yourself on the objectives.
   Read the whole LIFEPAC again.
   Recite through an oral examination.
   Review weak areas.
6. Take the LIFEPAC test as a closed book test.
7. LIFEPAC tests are administered and scored under direct teacher supervision. Students who receive scores below 80% should review the LIFEPAC using the SQ3R study method and take the Alternate Test located in the Teacher Handbook. The final test grade may be the grade on the Alternate Test or an average of the grades from the original LIFEPAC test and the Alternate Test.

GOAL SETTING AND SCHEDULES

Each school must develop its own schedule, because no single set of procedures will fit every situation. The following is an example of a daily schedule that includes the five LIFEPAC subjects as well as time slotted for special activities.

Possible Daily Schedule

8:15	-	8:25	Pledges, prayer, songs, devotions, etc.
8:25	-	9:10	Bible
9:10	-	9:55	Language Arts
9:55	-	10:15	Recess (juice break)
10:15	-	11:00	Math
11:00	-	11:45	History & Geography
11:45	-	12:30	Lunch, recess, quiet time
12:30	-	1:15	Science
1:15	-		Drill, remedial work, enrichment*

***Enrichment:** *Computer time, physical education, field trips, fun reading, games and puzzles, family business, hobbies, resource persons, guests, crafts, creative work, electives, music appreciation, projects.*

Basically, two factors need to be considered when assigning work to a student in the LIFEPAC curriculum.

The first is time. An average of 45 minutes should be devoted to each subject, each day. Remember, this is only an average. Because of extenuating circumstances a student may spend only 15 minutes on a subject one day and the next day spend 90 minutes on the same subject.

The second factor is the number of pages to be worked in each subject. A single LIFEPAC is designed to take 3 to 4 weeks to complete. Allowing about 3 to 4 days for LIFEPAC introduction, review, and tests, the student has approximately 15 days to complete the LIFEPAC pages. Simply take the number of pages in the LIFEPAC, divide it by 15 and you will have the number of pages that must be completed on a daily basis to keep the student on schedule. For example, a LIFEPAC containing 45 pages will require 3 completed pages per day. Again, this is only an average. While working a 45 page LIFEPAC, the student may complete only 1 page the first day if the text has a lot of activities or reports, but go on to complete 5 pages the next day.

Long-range planning requires some organization. Because the traditional school year originates in the early fall of one year and continues to late spring of the following year, a calendar should be devised that covers this period of time. Approximate beginning and completion dates can be noted on the calendar as well as special occasions such as holidays, vacations and birthdays. Since each LIFEPAC takes 3 to 4 weeks or eighteen days to complete, it should take about 180 school days to finish a set of ten LIFEPACs. Starting at the beginning school date, mark off eighteen school days on the calendar and that will become the targeted completion date for the first LIFEPAC. Continue marking the calendar until you have established dates for the remaining nine LIFEPACs making adjustments for previously noted holidays and vacations. If all five subjects are being used, the ten established target dates should be the same for the LIFEPACs in each subject.

SCIENCE PROJECTS LIST

Key

(1) = Those essential to perform for basic understanding of scientific principles.
(2) = Those which should be performed as time permits.
(3) = Those not essential for mastery of LIFEPACs.
S = Equipment needed for homeschool or Christian school lab.
E = Explanation or demonstration by instructor may replace student class lab work.
H = Suitable for homework or for homeschool students. (No lab equipment needed.)
V = This experiment is available on the Science Experiments video.

Science 401

401.A (2) H
401.B (1) H & V

Science 402

None

Science 403

403.A (1) H, S & V (seasonal)
403.B (2) H
403.C (3) S

Science 404

404.A (1) H
404.B (1) H & V
404.C (3) H
404.D (1) S & V
404.E (1) S & V
404.F (3) H & V
404.G (1) S
404.H (3) S & V

Science 405

405.A (1) S
405.B (1) S & V
405.C (2) S
405.D (1) S & V
405.E (1) S & V
405.F (3) S & V
405.G (2) S & V
405.H (1) S & V

Science 406

406.A (1) H & V
406.B (1) H & V
406.C (2) S & V
406.D (1) H
406.E (2) H & V
406.F (1) H & V
406.G (1) S & V
406.H (3) H & V
406.I (3) H

Science 407

407.A (1) H & V
407.B (2) H
407.C (1) H & V

Science 408

408.A (1) H
408.B (3) H
408.C (2) H & V

Science 409

409.A (1) H & V
409.B (1) H & V
409.C (2) S & V

Science 410

None

TEACHING SUPPLEMENTS

The sample weekly lesson plan and student grading sheet forms are included in this section as teacher support materials and may be duplicated at the convenience of the teacher.

The student grading sheet is provided for those who desire to follow the suggested guidelines for assignment of letter grades as previously discussed. The student's self test scores should be posted as percentage grades. When the LIFEPAC is completed the teacher should average the self test grades, multiply the average by .25 and post the points in the box marked self test points. The LIFEPAC percentage grade should be multiplied by .60 and posted. Next, the teacher should award and post points for written reports and oral work. A report may be any type of written work assigned to the student whether it is a LIFEPAC or additional learning activity. Oral work includes the student's ability to respond orally to questions which may or may not be related to LIFEPAC activities or any type of oral report assigned by the teacher. The points may then be totaled and a final grade entered along with the date that the LIFEPAC was completed.

The Student Record Book, which was specifically designed for use with the Alpha Omega curriculum, provides space to record weekly progress for one student over a nine-week period as well as a place to post self test and LIFEPAC scores. The Student Record Books are available through the current Alpha Omega catalog; however, unlike the enclosed forms, these books are not for duplication and should be purchased in sets of four to cover a full academic year.

WEEKLY LESSON PLANNER

Week of:

	Subject	Subject	Subject	Subject
Monday				
	Subject	Subject	Subject	Subject
Tuesday				
	Subject	Subject	Subject	Subject
Wednesday				
	Subject	Subject	Subject	Subject
Thursday				
	Subject	Subject	Subject	Subject
Friday				

WEEKLY LESSON PLANNER

Week of:

	Subject	Subject	Subject	Subject
Monday				
	Subject	Subject	Subject	Subject
Tuesday				
	Subject	Subject	Subject	Subject
Wednesday				
	Subject	Subject	Subject	Subject
Thursday				
	Subject	Subject	Subject	Subject
Friday				

Student Name ______________________ Year ____________

Bible

LP	Self Test Scores by Sections 1	2	3	4	5	Self Test Points	LIFEPAC Test	Oral Points	Report Points	Final Grade	Date
01											
02											
03											
04											
05											
06											
07											
08											
09											
10											

History & Geography

LP	Self Test Scores by Sections 1	2	3	4	5	Self Test Points	LIFEPAC Test	Oral Points	Report Points	Final Grade	Date
01											
02											
03											
04											
05											
06											
07											
08											
09											
10											

Language Arts

LP	Self Test Scores by Sections 1	2	3	4	5	Self Test Points	LIFEPAC Test	Oral Points	Report Points	Final Grade	Date
01											
02											
03											
04											
05											
06											
07											
08											
09											
10											

Student Name ______________________ Year ______________

Math

LP	Self Test Scores by Sections 1	2	3	4	5	Self Test Points	LIFEPAC Test	Oral Points	Report Points	Final Grade	Date
01											
02											
03											
04											
05											
06											
07											
08											
09											
10											

Science

LP	Self Test Scores by Sections 1	2	3	4	5	Self Test Points	LIFEPAC Test	Oral Points	Report Points	Final Grade	Date
01											
02											
03											
04											
05											
06											
07											
08											
09											
10											

Spelling/Electives

LP	Self Test Scores by Sections 1	2	3	4	5	Self Test Points	LIFEPAC Test	Oral Points	Report Points	Final Grade	Date
01											
02											
03											
04											
05											
06											
07											
08											
09											
10											

INSTRUCTIONS FOR SCIENCE

The LIFEPAC curriculum for grades two through twelve was written with the daily instructional material written directly in the LIFEPACs. The student is encouraged to read and follow his own instructional material, thus developing independent study habits. The teacher should introduce the LIFEPAC to the student, set a required completion schedule, complete Teacher checks, be available for questions regarding both subject content and procedures, administer and grade tests, and develop additional learning activities as desired. Teachers working with several students may schedule their time so that students are assigned a quiet work activity when it is necessary to spend instructional time with one particular student.

The Teacher Notes section of the Teacher's Guide lists the required or suggested materials for the LIFEPACs and provides additional learning activities for the students. The materials section refers only to LIFEPAC materials and does not include materials which may be needed for the additional activities. Additional learning activities provide a change from the daily school routine, encourage the student's interest in learning, and may be used as a reward for good study habits.

If you have limited facilities and are not able to perform all the experiments contained in the LIFEPAC curriculum, the Science Project List may be a useful tool for you. This list prioritizes experiments into three categories: those essential to perform, those which should be performed as time and facilities permit, and those not essential for mastery of LIFEPACs. Of course, for complete understanding of concepts and student participation in the curriculum, all experiments should be performed whenever practical. Materials for the experiments are shown in Teacher Notes - Materials Needed.

If you have limited facilities and are not able to perform all the experiments contained in the LIFEPAC curriculum, the Science Project List may be a useful tool for you. This list prioritizes experiments into three categories: those essential to perform, those which should be performed as time and facilities permit, and those not essential for mastery of LIFEPACs. Of course, for complete understanding of concepts and student participation in the curriculum, all experiments should be performed whenever practical. Materials for the experiments are shown in Teaching Notes—Materials Needed.

A suggested support item for this course is the 4th Grade Science Experiments video, SD0401. The video includes presentations of many of the experiments in this course. Several of the experiments that require special equipment or materials are demonstrated on these videos. They can either be used for answering the questions of the lab report or as a demonstration of the procedure prior to performing the experiment. A notice is included with each experiment in the LIFEPAC where the video is available.

SCIENCE 401

Unit 1: Plants

TEACHING NOTES

MATERIALS NEEDED FOR LIFEPAC	
Required	Suggested
• none	• fresh celery stalk (stem) with leaves on top • a glass of water • red ink or red food coloring • a knife • 4th Grade Science Experiments video

ADDITIONAL LEARNING ACTIVITIES

Section 1: Plant Life

1. Take a planned field trip to a supermarket. Divide into two groups. One group makes list of all vegetables; other group makes list of all fruits. Groups compare lists at school and discuss.
2. Small group working together makes list of all state flowers. Share list with class.
3. Memorize the first two verses of the hymn "How Great Thou Art." Write them from memory. Give what you have written to the teacher.
4. Draw a flower arrangement.
5. Using real or artificial flowers, make a floral arrangement and bring it to class.

Section 2: Plant Parts

1. Cut colored pictures of flowers and flower arrangements from magazines. Arrange bulletin board display.
2. Make leaf collection. Identify from reference books and share with class.
3. Bring seeds to class and classify them.
4. Plant some plants at home. Keep a diary of progress of the plants. Have one of your parents sign the report when you bring it to class.
5. Make a picture using various seeds.

ANSWER KEYS

SECTION 1

1.1 NL
1.2 NL
1.3 L
1.4 L
1.5 NL
1.6 NL
1.7 NL
1.8 L
1.9 NL
1.10 L
1.11 Teacher check
1.12 ____ RICK AND MARY TAKE A TRIP
1.13 _✓_ RICK AND MARY LEARN ABOUT FOOD
1.14 ____ RICK AND MARY HAD FRUIT FOR DESSERT
1.15 living
1.16 stem
1.17 bulb
1.18 root
1.19 food
1.20 e
1.21 c
1.22 a
1.23 b
1.24 d
1.25 Choices will vary.
1.26 Choices will vary.
1.27 Choices will vary.
1.28 Choices will vary.
1.29 Paragraphs will vary.
1.30 grape
1.31 apple
1.32 berry
1.33 walnut
1.34 cherry
1.35 orange
1.36 lemon or melon
1.37 Lists will vary.
1.38 Lists will vary.
1.39 Any order:
a. for food
b. for shelter
c. for enjoyment
d. for state symbols
1.40 vitamin C
1.41 trees
1.42 roots
1.43 Lists will vary.
1.44 Answers will vary.
1.45 forget-me-not
1.46 hibiscus
1.47 Example:
God loves us and wants us to have a beautiful world in which to live. God is good and would not have created a world less than beautiful.
1.48 Alaska
1.49 Choices will vary.

SELF TEST 1

1.01 water
1.02 grow
1.03 bulb
1.04 sugar
1.05 pepos
1.06 energy
1.07 pomes
1.08 vegetable
1.09 corn
1.010 South America
1.011 c. food
1.012 a. seeds
1.013 c. prune
1.014 c. leafy
1.015 b. bicycle
1.016 Any two of the following; Any order:
a. for food
b. for shelter
or for enjoyment
for state symbols
1.017 c
1.018 a
1.019 e
1.020 b
1.021 d
1.022 b
1.023 c
1.024 b

SECTION 2

2.1 flowers
2.2 leaves
2.3 roots
2.4 stems
2.5 below
2.6 seeds
2.7 flowers
2.8 a. flowers
b. leaves
c. stem
d. roots
2.9 Drawings will vary.
2.10 bark
2.11 bud
2.12 bud scales
2.13 future flower
2.14 future leaf
2.15 growing point
2.16 wood
2.17 a. Put red ink or coloring in a glass of water.
b. Set the celery stalk in the glass.
2.18 The red coloring should start to go up the celery stem.
2.19 Little tubes in the stem are red.
If the celery were examined under a microscope, the student could see little circles.
2.20 _____ a. The celery stem looked the same on the third day as it did on the first day.
__✓__ b. The water went up through the tubes of the celery plant.
_____ c. The celery stalk dried up.
2.21 swollen stem
2.22 above
2.23 the tubes in the stalk
2.24 Drawings will vary.
2.25 Blade
2.26 Stem
2.27 Hint:
Drawings could show a leaf blowing off a tree. The leaf will be green. The leaf will blow to the ground, and it will be brown as it lies on the ground. The leaf is decaying.
2.28 work
2.29 together
2.30 decay cycle
2.31 leaf
2.32 minerals
2.33 rain
2.34 near
2.35 deep
2.36 paint

2.37 season
2.38 green
2.39 roots
2.40 wheat
2.41 feed
2.42 road
2.43 seeds
2.44 ain
2.45 ee
2.46 oat
2.47 ear
2.48 ead
2.49 oast
2.50 ail
2.51 suf / fer
2.52 car / toon
2.53 in / vent
2.54 slen / der
2.55 un / kind
2.56 bap / tize
2.57 b. photosynthesis
2.58 c. happy
2.59 a. flowers
2.60 c. over 250,000
2.61 a. receptacle
2.62 b. stem
2.63 a. sepal
2.64 b. pollen
2.65 a. three
2.66 a. pistil
2.67 a. God
2.68 Choices will vary.
2.69 Choices will vary.
2.70 Choices will vary.
2.71 Choices will vary.
2.72 seeds
2.73 pome
2.74 rind
2.75 fruit

SELF TEST 2

2.01 c. leafy
2.02 a. enjoyment
2.03 b. run-ning
2.04 d. factory
2.05 b. citrus
2.06 c. stem
2.07 d. seeds
2.08 a. pepos
2.09 b. nonliving
2.010 a. starch
2.011 d. wood
2.012 a. bulb
2.013 true
2.014 false
2.015 true
2.016 true
2.017 true
2.018 false
2.019 false
2.020 true
2.021 true
2.022 true
2.023 true
2.024 false
2.025 false
2.026 true
2.027 false
2.028 true
2.029 true
2.030 false
2.031 true
2.032 true
2.033 false
2.034 false
2.035 false
2.036 true
2.037 true
2.038 true
2.039 Any four; any order:
bud, wood, bud scales, bark, future leaf, future flower, growing point
2.040 Any four; any order:
root, flowers (blossoms), leaves, stems
2.041 Examples:
fire, insects, disease, careless, cutting methods

LIFEPAC TEST

1. energy
2. bulb
3. nonliving
4. bud
5. berry
6. nut
7. citrus
8. minerals
9. flower
10. fire
11. d
12. f
13. e
14. a
15. c
16. b
17. true
18. false
19. false
20. true
21. false
22. true
23. true
24. Any order:
    a. root (system)
    b. stem
    c. leaf
    d. flower
25. honor
26. Lord's
27. decay

ALTERNATE LIFEPAC TEST

1. rind
2. root
3. oxygen
4. pollen
5. glory
6. enjoyment
7. d
8. h
9. g
10. e
11. b
12. f
13. c
14. i
15. a
16. true
17. false
18. true
19. false
20. true
21. false
22. true
23. true
24. true
25. true
26. Examples:
    fire
    over-cutting

SCIENCE 401

ALTERNATE LIFEPAC TEST

NAME ______________________________

DATE ______________________________

SCORE ______________________________

From the list of words, choose the correct word and write it in the blank (each answer, 4 points).

glory	enjoyment	pollen
rind	oxygen	root

1. Cucumbers are berries with a hard ____________ .
2. A parsnip is a vegetable with a large ____________ .
3. Plants give off ____________ into the air.
4. Bees pollinate flowers by carrying ______ from one flower to another.
5. The Psalmist wrote, "Thou art worthy, O Lord, to receive ____________ and honor and power."
6. Trees give us ____________ as well as shelter.

From the following list write the letter in each blank that makes the sentence correct (each answer, 4 points).

a. above ground	b. underground	c. flower
d. sugar	e. grow	f. stem
g. minerals	h. chlorophyll	i. leafy

7. We get energy from eating plants which contain ____________ .
8. The green coloring matter in plants that they use in making food is called ______________ .
9. Soil contains ____________ which plants use to make food.
10. God has given us everything we need to ____________ .
11. The white potato is a swollen stem which grows ____________ .
12. The flower of a plant is attached to the ____________ .

13. The stem of a plant contains the future ____________ .

14. Cabbage is a ____________ vegetable.

15. Bean seeds and pea seeds grow ____________ .

Write *true* or *false* (each answer, 3 points).

16. ______ The pioneers in America used corn for food.

17. ______ Spinach and cabbage are South American fruits.

18. ______ Each pistil of a flower is made up of three parts.

19. ______ The style of a flower is part of the pollen.

20. ______ Carbon dioxide is a gas.

21. ______ The celery stalk, which is eaten, grows underground.

22. ______ Herbs are plants whose stems and leaves are used in cooking.

23. ______ Apples and pears are fruit.

24. ______ A petal is a part of a flower.

25. ______ A melon is a fleshy fruit with a hard rind and many seeds.

Answer this question (each answer, 5 points).

26. Name two enemies of forests.

a. ______________________________ b. ______________________________

SCIENCE 402

Unit 2: Animals

TEACHING NOTES

MATERIALS NEEDED FOR LIFEPAC	
Required	Suggested
• none	• aquarium • fish • animal cage (small) • terrarium

ADDITIONAL LEARNING ACTIVITIES

Section 1: How Animals Are Structured

1. In late summer or early fall collect feathers of different birds. Observe colors and shapes of feathers. Look at different feathers under the microscope. Notice differences and likenesses. Make a display and share with the class.
2. Research all you can find about Jack Miner. Several students should work together and prepare a skit to share with the class.
3. Make a chart grouping animals into two groups: (a) animals that eat plants, (b) animals that eat other animals. Use as many pictures from magazines as possible.
4. Draw a map showing the routes taken by one or more of the following birds: Arctic tern, whistling swans, Canadian geese, wheatear.

Section 2: How Animals Act

1. Work with a friend. See how many different kinds of animals you can find mentioned in the Bible.
2. Research the topic: What Animals Hibernate? Report to the class.
3. Observe and report on the skills of a little kitten. Students who have one at home may wish to report on hunting skills.
   a. Notice how it plays with a rubber ball.
   b. Write down each movement of the kitten.
   c. What does it do with its front paws?
   d. What does it do with its hind feet?
   e. How does it hold its front paws when it pounces on a ball?
   f. Does it bite the ball?
   g. Are any of its movements similar to an adult cat catching a mouse?

4. Observe and report on the playing habits of a kitten.
    a. Attach a ball or some feathers to the end of a string.
    b. Dangle the ball or feathers above the kitten's head.
    c. How does it try to grab the ball?
    d. Let the kitten knock the ball under a chair or sofa. Watch how it reaches with one paw under the chair and retrieves the ball.

(Note: Instructor may wish to duplicate these instructions or similar ones for student use.)

Section 3: How Animals Are Provided For

1. A group of students may make a display for open house showing the different kinds of "houses" or homes that animals live in.
2. Several students may wish to work together on this activity. Make an audio recording of different sounds of animals—pets, animals in ponds, on farms. Play the sounds to small groups.
3. Build a birdhouse.
    a. Take a half gallon plastic bottle.
    b. Cut an entrance hole in it with a knife. Cut about half way up the side.
    c. Paint the outside a dark brown or dark green.
    d. Hang the birdhouse in a tree.
4. Build a self-designed birdhouse. Use scraps of wood from old boxes or pieces of small branches to look like a log cabin.
5. Make an insect collection. Secure directions from a good library book or online resources.
6. Study an interesting picture of children and pets, and write a story about the picture.
7. Write an original story about a pet you may have.

ANSWER KEYS

SECTION 1

1.1 strong
1.2 largest
1.3 mammal
1.4 two
1.5 tooth
1.6 ivory
1.7 false
1.8 true
1.9 true
1.10 false
1.11 true
1.12 true
1.13 false
1.14 true
1.15 b. warm-blooded
1.16 b. lungs and mouth or nose
1.17 a. blowhole
1.18 b. oil
1.19 false
1.20 true
1.21 false
1.22 false
1.23 false
1.24 true
1.25 true
1.26 false
1.27 false
1.28 true
1.29 Answers will vary.
The shape of the whale's spout or blow identifies the whale.

The blow of a blue whale is tall, slender and vertical, upwards of 9 meters in height. They are blue-gray with a blotchy appearance. The tail fluke is very broad with a straight or slightly concave bottom edge with a slight notch in the middle.

The blow of a right whale is a distinct V-shaped blow, upwards of 5 meters in height. They are mostly black with irregular white patches on the belly. The tail fluke is broad and symmetrical with a distinct notch.

The blow of a sperm whale is a low bushy blow, projected forward and to the left, usually less than 2 meters in height, very distinct. They are dark brownish gray with wrinkled skin. Each half of the tail fluke is the shape of a right triangle with a distinctive V notch in the middle. Sperm whales are easily identified by the large head.

1.30 c. follow a planned course
1.31 a. timetable of coming and going
1.32 d. bird
1.33 b. Arctic tern
1.34 nests
1.35 flocks
1.36 seeds
1.37 breeding
1.38 three thousand (or 3,000)
1.39 short
1.40 runway
1.41 South (or Carolinas)
1.42 nests
1.43 V
1.44 c
1.45 a
1.46 e
1.47 f
1.48 b
1.49 d
1.50 reptile
1.51 snake
1.52 armor
1.53 crawl
1.54 legs
1.55 water
1.56 cold
1.57 snake (or reptile)
1.58 legs
1.59 Cold
1.60 Examples; any order:
a. lizards
b. snakes
c. turtles and tortoises
d. crocodiles
1.61 Example:
God created the fish with fins, a body that moves, and gills with which to breathe.
1.62 Example:
The fish swims by wiggling its body and using its tail fins.
1.63 Example:
The fish has two pairs of fins in place of legs and arms. The fins act as limbs in helping the fish travel. The fins are like broad paddles which drive the fish through the water.

1.64 Example:
The fish turns by using its tail and body.
1.65 A fish's scales overlap, one on top of another.
1.66 Teacher check
1.67 To its stomach
1.68 Examples:
lion, tiger, leopard
1.69 Members of the cat family have sharp teeth because they are meat-eaters. (or because they tear their meat from the bone before swallowing it)
1.70 Sheep have no cutting teeth on the upper jaw. Sheep's teeth are made for eating grass.
1.71 breathing (or taking in air)
1.72 tubes
1.73 nose
1.74 lungs
1.75 lungs
1.76 sacs
1.77 air (oxygen)
1.78 gills
1.79 oxygen
1.80 body

SELF TEST 1

1.01 d
1.02 f
1.03 e
1.04 b
1.05 c
1.06 h
1.07 a
1.08 g
1.09 true
1.010 true
1.011 false
1.012 false
1.013 false
1.014 true
1.015 true
1.016 true
1.017 true
1.018 true
1.019 true
1.020 false
1.021 false
1.022 true
1.023 false
1.024 true
1.025 false
1.026 true
1.027 true
1.028 false
1.029 false
1.030 true
1.031 false
1.032 false
1.033 c. herd
1.034 b. South
1.035 a. follow a planned course
1.036 c. much different
1.037 b. meat-eaters
1.038 a. bird refuge
1.039 a. Arctic Tern
1.040 b. backbones
1.041 Examples; any order:
a. lizard
b. snake
c. turtles and tortoises
d. crocodiles
1.042 Examples: any four in any order:
bear live young; nurse their young; have hair; breathe with lungs; have backbones; are warm-blooded

1.043 Examples; any order:
They build their nests in Arctic regions. They need a long runway to get into the air. They fly in a V pattern. They eat worms, shell fish, seeds, and roots.

SECTION 2

2.1 true
2.2 true
2.3 false
2.4 false
2.5 true
2.6 true
2.7 false
2.8 true
2.9 e. a stream
2.10 c. 1,000
2.11 i. two
2.12 b. one
2.13 d. five
2.14 k. waterfalls
2.15 h. fifteen
2.16 f. difficult
2.17 a. ocean
2.18 j. hatched
2.19 g. 30,000
2.20 true
2.21 false
2.22 true
2.23 true
2.24 false
2.25 Examples; any order:
size of territory the bird needs success in driving off rivals amount of food available
2.26 Teacher check
2.27 Hint:
The first sentence should be a topic sentence.
2.28 Teacher check
2.29 Teacher check
2.30 Teacher check
2.31 Teacher check
2.32 ti / ger
2.33 na / ture
2.34 lo / cate
2.35 mo / ment
2.36 mo / tel
2.37 trav / el
2.38 riv / er
2.39 hab / its
2.40 tim / id
2.41 took
2.42 boot
2.43 bought
2.44 owl
2.45 Example:
The owl hooted in the tree.
2.46 Example:
The boy shook the apple tree.

2.47 Example:
Milk will spoil if it is not refrigerated.
2.48 Example:
Some people pout if they do not get their way.
2.49 chirps
2.50 thirty
2.51 temperature

SELF TEST 2

2.01 warm-blooded
2.02 extinct
2.03 cold-blooded
2.04 Bible
2.05 rule
2.06 protect
2.07 spawning ground
2.08 weaverbird
2.09 territory
2.010 ivory
2.011 false
2.012 true
2.013 true
2.014 true
2.015 false
2.016 true
2.017 false
2.018 true
2.019 true
2.020 false
2.021 true
2.022 true
2.023 false
2.024 true
2.025 false
2.026 true
2.027 false
2.028 false
2.029 Count the number of chirps in fifteen seconds then add thirty.
2.030 Example:
Most workers leave the hive. Several workers stay at the entrance and fan their wings very fast. This fanning drives the air into the hive.
2.031 whale
2.032 blow-hole
2.033 shrew
2.034 scraper
2.035 lings
2.036 dance

SECTION 3

3.1 c
3.2 a
3.3 a
3.4 b
3.5 a
3.6 c
3.7 b
3.8 true
3.9 false
3.10 false
3.11 true
3.12 false
3.13 false
3.14 true
3.15 You should have a check mark by: a, c, d, g, h, i, j.
3.16 true
3.17 true
3.18 false
3.19 true
3.20 false
3.21 true

SELF TEST 3

3.01 snakes
3.02 radar
3.03 oxygen
3.04 hollow
3.05 meat
3.06 tubes
3.07 home
3.08 reptiles
3.09 false
3.010 false
3.011 false
3.012 true
3.013 false
3.014 true
3.015 sanctuary (refuge)
3.016 Miner
3.017 Audubon
3.018 God's
3.019 follow
3.020 grass
3.021 gills
3.022 Arctic tern
3.023 instinct
3.024 bat
3.025 a. reptile
3.026 c. leg
3.027 b. cold-blooded
3.028 a. rudder
3.029 a. lungs
3.030 b. lungs
3.031 b. refuge
3.032 c. instinct

LIFEPAC TEST

1. i
2. g
3. l
4. a
5. f
6. h
7. m
8. c
9. e
10. b
11. d
12. j
13. false
14. true
15. true
16. false
17. true
18. true
19. true
20. true
21. false
22. false
23. false
24. c. queen
25. c. in herds
26. b. sharp
27. a. hollow
28. c. gills
29. c. provided by God
30. Any order:
    queen, drone, workers
31. Count the number of chirps in 15 seconds.
    Add 30.
32. Bats have poor eyes or eyesight, can fly by instinct, can detect the faintest sound, and can send out high-pitched sounds (squeaks).
33. Example:
    He created them.
    He supplies food for them.
    He supplies shelter.
    He has given animals protective coloration.
    He has given animals instincts.
34. Examples:
    Man is trying to correct his mistakes in not taking care of wildlife.
    Groups of people are working hard to protect animals.
35. Examples:
    It swims far away from where it was born.
    When it is an adult, it finds its way back to its birthplace.

ALTERNATE LIFEPAC TEST

1. d
2. f
3. b
4. j
5. a
6. i
7. e
8. k
9. c
10. g
11. true
12. false
13. false
14. true
15. false
16. true
17. true
18. true
19. true
20. true
21. a. south
22. b. refuge
23. c. Canada
24. a. spawning grounds
25. a. night
26. a. migrating
27. b. reptiles
28. c. buffalo
29. c. a thousand
30. a. elephant
31. backbone
32. elephant
33. warm
34. snake
35. a. queen
    b. workers
    c. drones
36. Count the number of chirps for 15 seconds, then add 30.
37. Example:
    for food to serve man
38. Example:
    It sends out signals and receives them again.

SCIENCE 402

ALTERNATE LIFEPAC TEST

NAME ________________________________

DATE ________________________________

SCORE ________________________________

Match these items (each answer, 2 points).

1. _______ ostrich
2. _______ whale
3. _______ inborn
4. _______ hive
5. _______ wheatear
6. _______ reptile
7. _______ meat-eater
8. _______ largest brain
9. _______ ivory
10. _______ shrew

a. navigates
b. instinct
c. tusk
d. largest bird
e. bear
f. largest mammal
g. smallest mammal
h. large plant
i. turtle
j. honeybee's home
k. sperm whale

Write *true* or *false* (each answer, 2 points).

11. _______ The bones of most birds are hollow.

12. _______ The ostrich is a swift-flying bird.

13. _______ In the wilds of Africa, elephants usually travel alone.

14. _______ A young salmon is called a smolt.

15. _______ The male honeybee is called a worker.

16. _______ Meat-eating animals have sharp teeth.

17. _______ The gills of a fish help it to breathe.

18. _______ Ivory is secured from the tusks of elephants.

19. _______ Some birds migrate each year.

20. _______ Whales are not fish.

Write the letter of the correct answer on each line (each answer, 2 points).

21. Every year in the winter, flocks of swans fly ______________________ .
a. south b. north c. west

22. A place where birds or mammals are protected is called a __________ .
a. nest b. refuge c. lake

23. Jack Miner's bird refuge was located in __________________________ .
a. South Carolina b. the Arctic c. Canada

24. The adult salmon swims upstream to its old ______________________ .
a. spawning grounds
b. waterfall
c. ocean feeding grounds

25. Bats usually fly at ____________________________ .
a. night b. noon c. eleven o'clock

26. When geese fly south to spend the winter in a warmer climate, it is known as ______________ ______________________ .
a. migrating b. renewing c. flocking

27. Snakes belong to an animal class known as ____________________ .
a. mammals b. reptiles c. amphibians

28. Man has cruelly and needlessly killed most of the world's __________ , that once were numbered in the millions.
a. horses b. bears c. buffalo

29. The number of eggs a female salmon can lay at one time is about ____ .

a. 2 b. 12 c. a thousand

30. The largest of all land animals is the ______________ .

a. elephant b. giraffe c. black bear

Write the correct word in the blank (each answer, 4 points).

31. A vertebrate is an animal that has a ____________________________ .

32. Ivory comes from the teeth (called "tusks") of the _________________ .

33. All mammals have a _____________________ body temperature.

34. A python is a ____________________________ .

Answer this question (each answer, 3 points).

35. What three kinds of honey bees live in the hive?

a. ____________________________________

b. ____________________________________

c. ____________________________________

Answer these questions (each numbered item, 5 points).

36. How would you find the temperature of the air if you heard a tree cricket chirping?

____

____

37. What is God's purpose for animals?

____

____

38. How does a bat travel in the dark without hitting anything?

____

____

SCIENCE 403

Unit 3: Man and His Environment

TEACHING NOTES

MATERIALS NEEDED FOR LIFEPAC	
Required	Suggested
• one or more kitchen strainers with handles lengthened by attaching them to long sticks or broom handles • a number of wide-mouthed jars—one-liter, two-liter and four-liter sizes • pans • an aquarium if one is available • a hand magnifying glass • some cloth or netting to put over jars • some string • a microscope (for classroom use)	• 4th Grade Science Experiments video

ADDITIONAL LEARNING ACTIVITIES

Section 1: Man Depends on God's Plan

1. Visit a farm with other students. Ask questions of the farmer about types of soil, farm equipment, methods of plowing, methods of harvesting, and other things covered in this LIFEPAC.
2. Other field trips might be made to the county agricultural agent, forest ranger tower or station, or a wildlife preserve.
3. Put some algae (green scum) from a pond into two jars of pond water. Place one in a sunny spot on the windowsill and one in a closet. Compare the growth and report to the class.
4. Student may do special research on photosynthesis and report to class.

Section 2: Man Depends on Communities

1. Ask students who have been to the seashore to tell about their experiences with different types of sea life.
2. Read a story about animals to the class and discuss it.
3. Have a good reader select an exciting story about insects or reptiles and read it to the class. Discuss details with the class.
4. If the class makes or has a desert terrarium, several students may be assigned to write daily observations and notes regarding the activities in it.
5. Several students may be permitted to go outside during class sessions and study an ant hill. They could report to class some of their observations.
6. Select four or five animals and write a short description of their homes.
7. Do some research on animal "languages" and communication. For example, find out how bees and dolphins communicate. (The out of print book, The Language of Animals, by Millicent Ellis Selsam is one possible resource.)

Section 3: Man Tries to Save His Environment

1. With a friend visit a museum that has a display on conservation. Take notes and share them with the class.
2. Get together with a few classmates. Read Genesis 1:28 from the Bible and discuss it.
3. Students may wish to research the subject of "State Forests" and write a paper on the subject.
4. One or more students may wish to look up information on National Monuments and report to the class. Ask them to find out why certain monuments were selected by our government.
5. A student interested in photography may take snapshots of pollution (refuse in streams, unsightly dumps, etc.) and show to the class.
6. Student may research "Solar Energy" and tell the class some of its advantages.

ANSWER KEYS

SECTION 1

1.1 Study of the home
1.2 Ecology is the science that deals with living things and how they are related to each other and their environments.
1.3 Example:
I eat cherries from a tree.
1.4 Example:
I drink milk from a cow.
1.5 Answers will vary.
1.6 Answers will vary.
1.7 The machines would dig up his home.
1.8 Example:
He would have neighbors.
1.9 Ecology is the study of all the life God put on earth.
1.10 Environment is the place or general area where you live.
1.11 Ecologists
1.12 the connection of all life on earth
1.13 Examples:
Drinking, swimming, watering lawns, washing clothes, and others
1.14 Example:
The effects of the water do not reach far inland.
1.15 air
1.16 sun
1.17 Either order:
a. nitrogen
b. oxygen
1.18 vapor
1.19 blanket
1.20 water
1.21 heat
1.22 God
1.23 the sun
1.24 light
1.25 a. light
b. air
c. water
d. soil
1.26 Teacher check
1.27 you-peach-sun
1.28 you-cow-grass-sun
1.29 you-milk-cow-grass-sun
1.30 boy-cow-grass-sun
1.31 cat-bird-bug-plant-sun
1.32 hawk-snake-frog-dragonfly-water-plant-sun
1.33 cat-mouse-grass-sun
1.34 Teacher check
1.35 predator
1.36 consumer
1.37 producer
1.38 predator
1.39 predator
1.40 consumer
1.41 producer
1.42 Teacher check
1.43 decomposers
1.44 pests
1.45 bacteria
1.46 Either order:
a. bacteria
b. mold
1.47 termite
1.48 Example:
In 1850 three pairs of rabbits were taken from Europe to Australia and turned loose. Within a few years they increased so rapidly that they destroyed food crops. Australia spent millions of dollars to get rid of the rabbits in order to protect the food supply for humans.
1.49 producers, consumers, decomposers
1.50 producers make food, consumers eat plants or animals, decomposers change waste back to nutrients for the soil
1.51 photosynthesis
1.52 The cells in the plant leaves combine energy from the sun with water, minerals and carbon dioxide to make food.
1.53 water, air, soil, and light
1.54 Example:
There must be just the right number of plants and animals in a community for all to be able to live.

SELF TEST 1

1.01 d
1.02 f
1.03 a
1.04 i
1.05 h
1.06 g
1.07 c
1.08 j
1.09 e
1.010 b
1.011 ecologist
1.012 population
1.013 mold (or bacteria)
1.014 sun
1.015 communities
1.016 food chain
1.017 Any order:
a. air
b. water
c. light
d. soil
1.018 bacteria
1.019 water
1.020 God
1.021 grasshopper
1.022 birds
1.023 bears
1.024 fungi
1.025 clover
1.026 water lily
1.027 rabbit
1.028 frog
1.029 cow
1.030 mold
1.031 cat
1.032 spider
1.033 snake
1.034 termite
1.035 orange tree
1.036 heat
1.037 sun
1.038 energy
1.039 The right number of plants and animals are in a community.
1.040 All life is connected.
1.041 Consumers eat many times their weights in food during their lifetimes.

SECTION 2

2.1 Teacher check
2.2 Teacher check
2.3 Teacher check
2.4 Describe beads, rods, knots, mats, and flat networks.
2.5 Example:
Top-fish, frogs, Middle-minnows, tadpoles, wigglers, Bottom-algae, water plants
2.6 Example:
alligator-bass-tadpoles-water-bugs, algae-sun
2.7 Teacher check
2.8 true
2.9 false
2.10 false
2.11 true
2.12 living
2.13 Either order:
a. plant
b. animal
2.14 crayfish
2.15 beaver
2.16 d
2.17 a
2.18 b
2.19 Teacher check
2.20 Teacher check
2.21 Some desert plants have very long roots; others keep their moisture instead of putting it into the air; some make food in their stems. All survive with very little moisture.
2.22 Example:
fox-rabbit-plants-sun
2.23 Examples:
a. mesquite
b. grasshopper
c. elf owl
d. vulture
2.24 Example:
fox-squirrel-berries-sun
2.25 a. top level trees
b. middle level plants
c. lower level (ferns and berries)
2.26 Teacher check
2.27 men, women, boys, girls, dogs, cats, flowers, shrubs, garden vegetables, grasshoppers, and so forth
2.28 parents
2.29 Example:
grocer
2.30 Examples:
farmer, rancher
2.31 parents
2.32 store
2.33 Examples:
farmer, manufacturer
2.34 He raises it.
2.35 make food, photosynthesis
2.36 sun
2.37 food chain
2.38 yes
2.39 Examples:
pets, milk, meat, leather, clothing
2.40 yes
2.41 Examples:
Animals depend upon humans in the following ways: pets for love and food, domestic animals such as cows for food and care wild animals and plants for protection and wise use
2.42 Humans were to learn about animals and plants and how to care for them.
2.43 Examples:
prayer, Bible study
2.44 Examples:
home, parents, family, friends
2.45 Examples:
schools, libraries, museums
2.46 language
2.47 tools
2.48 Any order:
a. ocean
b. still water
c. river
d. field
e. desert
f. forest
g. human
2.49 Any order:
a. all include plants and animals
b. all the members are dependent on each other
c. all are made up of producers, consumers, decomposers
d. all need air, water, soil, and light to live
2.50 The answers to 2.49 are all true for the human community.
2.51 Man has needs that plants and animals do not have.

2.52 Any order in list:
a. count
b. spout
c. blouse
d. scout
e. how
f. frown
g. owl
h. shower
i. flower
j. throw
k. tomorrow
l. blower
m. show
n. yellow
o. slower
p. borrow
q. shadow
r. sorrow

SELF TEST 2

2.01 habitat
2.02 predator
2.03 producer
2.04 communities
2.05 plankton
2.06 photosynthesis
2.07 e
2.08 g
2.09 a
2.010 c
2.011 f
2.012 b
2.013 i
2.014 d
2.015 owl or elf owl
2.016 roots
2.017 cutting down
2.018 image of God
2.019 Any order:
a. strong swimmers
b. animal life on the ocean floor
c. plankton
2.020 Any order:
a. producer
b. consumer
c. decomposer
2.021 Any order:
a. water
b. air
c. light
d. soil
2.022 Examples:
a. jack rabbits
b. elf owls
2.023 Examples:
a. otter
b. beaver
2.024 Examples:
a. rabbit
b. prairie dog
2.025 Examples:
field community-rabbit-grass-sun
2.026 Any order:
a. top level
b. middle level
c. lower level

SECTION 3

3.1 Teacher check
3.2 erosion
3.3 pollution
3.4 extinct
3.5 Teacher check
3.6 Teacher check
3.7 Teacher check
3.8
a. bar / ber
b. mar / gin
c. or / phan
d. nor / mal
e. tur / key
f. har / ness
3.9
a. daugh / ter
b. boi / ler
c. oy / ster
d. coun / ter
e. au / tumn
f. point / ing
3.10
a. kitch / en
b. teach / er
c. brush / es
d. ditch / es
e. preach / er
f. weath / er
3.11 Teacher check
3.12 c
3.13 d
3.14 a
3.15 f
3.16 e
3.17 Teacher check
3.18 largest underground cave in the world
3.19 the only tropical area in the United States
3.20 very large canyon through which the Colorado River flows
3.21 the top of a huge dormant (sleeping or quiet) volcano
3.22 pueblos or houses of prehistoric cliff dwellers
3.23 highest mountain in the United States
3.24 six forests of petrified wood, prehistoric Indian dwelling and Painted Desert
3.25 largest and oldest living things in the world
3.26 geysers, hot springs, waterfalls, mountains
3.27 river canyon of huge rock formations
3.28 wearing away of the soil
3.29 Examples:
by planting trees, contour farming
3.30 waste and dirt
3.31 Examples:
stop gasoline exhaust, stop factory smoke, stop burning trash
3.32 Examples:
stop factory and sewage dumping
3.33 solar, nuclear
3.34 I would use it wisely.
3.35 I would keep it safe.

SELF TEST 3

3.01 i
3.02 e
3.03 f
3.04 b
3.05 h
3.06 c
3.07 j
3.08 g
3.09 a
3.010 d
3.011 a. leaving soil loose with no cover crop
b. planting the same crop in the same field over and over, cutting forests
3.012 a. sewage, factory waste, pesticides, using up ground
b. water, litter, and rubbish thrown in water
3.013 Examples:
a. smoke
b. car exhaust or factory waste materials
3.014 a. otter
b. beaver
3.015 b
3.016 a
3.017 b
3.018 c
3.019 b
3.020 c
3.021 c
3.022 b
3.023 I would keep it safe.
3.024 I would use it wisely.
3.025 Any order:
a. water
b. light
c. air
d. soil
3.026 a. producer
b. consumer
c. decomposer
3.027 whale-fish-plankton-sun
3.028 sun
3.029 "web of life"

LIFEPAC TEST

1. e
2. j
3. a
4. h
5. c
6. f
7. b
8. d
9. g
10. i
11. b
12. c
13. a
14. a
15. a
16. b
17. b
18. c
19. a
20. b
21. a. strong swimmers
    b. all water animals living on the ocean floor
    c. plankton
22. Any order:
    a. air
    b. water
    c. light
    d. soil
23. Examples; either order:
    a. he has littered
    b. he has cut down too many trees
24. a. top level
    b. middle level
    c. lower level
25. you-milk-grass-sun (cow)
26. false
27. false
28. true
29. true
30. false
31. true
32. false
33. true
34. Animals feed on plants. Other animals feed on those animals. Just the right number of living things are in a community.

ALTERNATE LIFEPAC TEST

1. g
2. a
3. d
4. i
5. b
6. j
7. f
8. e
9. c
10. h
11. b
12. a
13. b
14. a
15. b
16. c
17. a
18. c
19. b
20. c
21. Example:
    I can stop litter, write letters, and stop waste.
22. Example:
    God gave me the responsibility to learn about living things and care for them.
23. Example:
    Animals and birds migrate to escape cold weather or perhaps to secure food. No one knows.
24. Any order:
    a. top
    b. middle
    c. lower
25. Example:
    a. littered
    b. cut down too many trees
26. Examples; any order
    a. air
    b. soil
    c. light or water or minerals
27. a. strong swimmers
    b. ocean floor life
28. true
29. true
30. false
31. false
32. false

SCIENCE 403

ALTERNATE LIFEPAC TEST

NAME ______________________

DATE ______________________

SCORE ______________________

Match these items (each answer, 2 points).

1.	_______	pollution	a.	to use wisely
2.	_______	conserve	b.	sun energy
3.	_______	life cycle	c.	air surrounding earth
4.	_______	erosion	d.	changes during life
5.	_______	solar	e.	how plants make food
6.	_______	rot	f.	decomposer
7.	_______	crayfish	g.	waste, dirt
8.	_______	photosynthesis	h.	killed off
9.	_______	atmosphere	i.	loss of top soil
10.	_______	extinct	j.	to decay

Write the correct letter on each line (each answer, 3 points).

11. When water is put on dry soil, the bubbles that appear show that the soil contains ______________________ .

a. water b. air c. nutrients

12. Mold is a ______________________ .

a. decomposer b. pest c. predator

13. Replanting forests ______________________ .

a. has no effect on our water supply
b. helps return water to underground reservoirs
c. is a waster of young trees

14. Plants and animals depend on each other for ____________ .
a. food b. light c. both a and b

15. Trees provide homes for ____________ .
a. deer b. birds c. foxes

16. Nature's food makers are ____________ .
a. plants b. trees c. both a and b

17. Conserve means to ____________ .
a. protect from loss b. waste c. wait

18. Animals and plants live together in ____________ .
a. houses b. habitats c. communities

19. Animals that eat other animals are called ____________ .
a. producer b. predators c. neither of these

20. The world we live in ____________ .
a. never changes b. has stopped changing c. changes all the time

Answer these questions in sentences (each answer, 5 points).

21. What are some things you can do to help conserve and clean up your environment?

____

____

22. What is the responsibility God gave you in Genesis 1:28?____________

____

____

23. Why do animals and birds migrate? ____________

____

Answer these questions (each answer, 3 points).

24. What are the three levels of a forest?

a. ____________

b. ____________

c. ____________

25. What are two ways that man has been careless with his environment?

a. ____________

b. ____________

26. What are three natural resources God has provided for us?

a. ______________________________

b. ______________________________

c. ______________________________

27. What are two of the groups of water animals in the ocean?

a. ______________________________

b. ______________________________

Write *true* or *false* (each answer, 1 point).

28. ____________ A clean countryside attracts animals.

29. ____________ God made a perfect environment for living things by giving resources.

30. ____________ Nearly all living things live alone.

31. ____________ A rabbit is a producer.

32. ____________ Pollution is good for most people.

SCIENCE 404

Unit 4: Machines

TEACHING NOTES

MATERIALS NEEDED FOR LIFEPAC	
Required	**Suggested**
• one sheet of poster board (any color) • one meter stick or yard stick • two felt tip pens (black and red) • one sharp pencil with an eraser • a large nail, a hammer • a block of wood • four textbooks, each at least 3 centimeters thick • a toy truck or car • a board about 1 meter long that is wider than the toy truck • a spring balance • one wood screw and one nail, each about 2 1/2 centimeters long • two thin pieces of wood • one screwdriver • a drawing compass • a spring balance • heavy twine • corrugated cardboard (from sides of a heavy box) • a piece of wood about 60 centimeters long and 30 centimeters wide • a broomstick cut to the length of 60 centimeters • two wood side pieces (2x2's would be best) at least 30 centimeters long, each notched on one end so the broomstick will lie loosely in the notches • two pulleys • a light rope about two meters long • a two-kilogram weight • a spring scale • a broomstick	• one seesaw and a friend • one wood screw and one nail each about 2 1/2 centimeters long • two thin pieces of wood • one hammer • one screwdriver • 4th Grade Science Experiments video

ADDITIONAL LEARNING ACTIVITIES

Section 1: Machines are Needed

1. Place a board at a slight slope. Let a toy truck roll down the slope. Now place it at the top of the slope and place a small block of wood in front of each wheel. Allow several students to try it. Discuss with the students the reason why the blocks cling to the board and prevent the truck from rolling down the slope.
2. Have the class experiment with weights. Lift a one-kilogram weight with one hand and a two-kilogram weight with the other. Use other weights and explain why it is harder to lift the heavier weights.
3. Put several books in a cardboard box and push it a short distance across a table top. Put several pencils parallel to each other and in a straight line so the box will roll on them. Push the box again. How do the pencils allow the box to move more easily? Explain to the class.
4. Bring a bicycle wheel to class and explain how the wheel bearings make it easier for the wheel to turn.
5. Find a way to demonstrate that oil or grease help reduce friction in machines and wheels.

Section 2: Machines are Simple

1. Bring a variety of wedges to class. Demonstrate each one and discuss with the class how they are used. (Examples: nail, chisel, axe head, knife, pins, and needles.)
2. Collect pictures from picture file or magazines of objects mentioned in the word list Section 2. Give students cards with words printed on them. Show picture and have student who holds correct card hold it up. (Choose sides and make a game of it.)
3. Have students make a collection of levers (tweezers, nutcracker, hammer, hand can opener, etc.). Each child should determine the work that each tool does and tell how it makes that particular work easier.
4. Secure two single fixed pulleys to two posts in the school yard. Connect them with a continuous rope. Let children experiment with moving a load across the distance between the posts by fastening the load to the rope and operating the pulleys.
5. Student may select a thick board and bring hammer to class. He drives a large nail a short way into the board and demonstrates and explains how it is easier to draw the nail out of the board using a hammer that it would be to pull it out by hand. Other students may try pulling it out by hand.
6. Student may write questions about levers, inclined planes, wedges, wheels and axles, and pulleys for class discussion after Section 2 is completed.

Section 3: Machines are Complex

1. Group work together on building several model machines (cars, boats, etc.) from model kits purchased from local toy stores or hobby shops.
2. Consult reference books or online resources and find out how the great pyramids of Egypt were built without cranes and derricks. Give a report contrasting methods of lifting heavy stones today.
3. Make a model of an ancient catapult similar to that mentioned in 2 Chronicles 26:15.
4. Draw a picture of a catapult being used in battle.

ANSWER KEYS

SECTION 1

1.1 Experimenting is the best way to get scientific information. Experiments are valuable only when closely and accurately observed.
1.2 He experimented to prove the pull of gravity. It is said he dropped objects of different weights from a tower to prove they would reach the ground at the same time.
1.3 book
1.4 sheet of paper
1.5 yes
1.6 gravity pulls them equally
1.7 Rounded wings and bodies cut down air resistance and decrease the effect of friction so that the plane can fly at a high speed.
1.8 yes
1.9 The spirit of God hath made me, and the breath of the Almighty hath given me life.
1.10 yes
1.11 when I am sleeping
1.12 when I am moving around
1.13 My hands feel warm.
1.14 heat
1.15 Teacher check
1.16 a. jawbone
b. saw
c. automobile
d. autumn
e. daughter
f. yawn
g. caught
h. taught
i. bought
j. fought
k. raw
l. lawful
m. haunted
n. nautical
o. naughty
1.17 a. 3
b. 1
c. 1
d. 1
e. 3
f. 3
g. 1
h. 1
1.18 b, c, d

SELF TEST 1

1.01 pushing or pulling something to move it
1.02 Stored energy is energy at rest. Energy in action is energy being used to move something.
1.03 from the sun
1.04 machines
1.05 gravity
1.06 friction
1.07 energy
1.08 earth
1.09 force
1.010 energy in action
1.011 Galileo
1.012 James Watt
1.013 King Solomon
1.014-1.017 Any order:
1.014 heat
1.015 light
1.016 sound
1.017 electricity
1.018 They would have used friction and gravity.
1.019 Energy is the ability to do work.
1.020 Man needs simple machines to help him do work easier.
1.021 d
1.022 a
1.023 c or h
1.024 c or h
1.025 f
1.026 e
1.027 i
1.028 b
1.029 g

SECTION 2

2.1 Teacher check
2.2 hammer
2.3 The hammer is a lever used to increase force.
2.4 easier
2.5 yes
2.6 less
2.7 Teacher check
2.8 Teacher check
2.9 Teacher check
2.10 pull it up inclined plane
2.11 because more distance makes less effort necessary
2.12 Teacher check
2.13 Teacher check
2.14 Teacher check
2.15 Teacher check
2.16 Abraham Lincoln-rail splitter
2.17 Teacher check
2.18 Teacher check
2.19 wider
2.20 felt tip pen
2.21 The pitch is wider. The wider the pitch, the faster the screw can move up and down when force or effort is applied.
2.22 The screw won't pull out of the wood easily.
2.23 The wood screw is harder to pull apart.
2.24 The screw or inclined plane has twisted into the wood.
2.25 screw
2.26 yes
2.27 A screw fastens things together more tightly than a nail.
2.28 Teacher check
2.29 Teacher check
2.30 Teacher check
2.31 The distance the object was lifted was the same each time.
2.32 The string is exactly as long as the axle is round.
2.33 less
2.34 Teacher check
2.35 friction
2.36 wheel and axle
2.37 less
2.38 gears
2.39 speed
2.40 A wheel is a roller. A gear is notched.
2.41 The chain fits into the teeth of the two wheels. As you pedal, the larger wheel turns the smaller wheel.
2.42 Teacher check
2.43 2 kg
2.44 no
2.45 1 kg
2.46 The weight is lifted by 2 strands of rope.
2.47 2
2.48 Teacher check

SELF TEST 2

2.01-2.06 Any order:
2.01 wedge
2.02 inclined plane
2.03 wheel and axle
2.04 lever
2.05 screw
2.06 pulley
2.07 a
2.08 d
2.09 b
2.010 a
2.011 e
2.012 c
2.013 f
2.014 d
2.015 c
2.016 e
2.017 c
2.018 fulcrum
2.019 less
2.020 sloping
2.021 force
2.022 lever
2.023 inclined plane
2.024 wedge
2.025 friction
2.026 axle
2.027 force
2.028 gears
2.029 direction
2.030 less
2.031 advantage
2.032 a. work
b. force
2.033 thread
2.034 pitch
2.035 spiral
2.036 simple machines

SECTION 3

3.1 Examples:
lawn mower, vacuum cleaner, washing machine
3.2 Examples:
car, bicycle, truck, train
3.3 Teacher check
3.4 Teacher check
3.5 Teacher check
3.6 Teacher check
3.7 yes
3.8 Examples:
heaters, air conditioners, fans
3.9 Teacher check
3.10 Teacher check
3.11 Teacher check
3.12 yes
3.13 Examples:
trucks, milk machines, packaging machines
3.14 Teacher check
3.15 Teacher check
3.16 Teacher check

SELF TEST 3

3.01 false
3.02 true
3.03 true
3.04 false
3.05 true
3.06 false
3.07 false
3.08 true
3.09 force
3.010 complex
3.011 complex
3.012 Work is pushing or pulling something to move it.
3.013 Gravity is an attraction between two objects that pulls things towards the center of the earth.
3.014 Friction is the force present when two things rub together.
3.015 Example:
A boy standing still is an example of stored energy.
3.016 Example: A boy running is an example of energy in action.
3.017 Either order:
a. gravity
b. friction
3.018 energy
3.019 Galileo
3.020 lever
3.021 machine
3.022-3.027 Any order; example:
3.022 lever — hammer
3.023 inclined plane — ramp
3.024 wedge — axe
3.025 wheel and axle — doorknob
3.026 pulley — block and tackle
3.027 screw — wood screw

LIFEPAC TEST

1. d
2. f
3. i
4. j
5. a
6. h
7. c
8. g
9. b
10. e
11. false
12. true
13. false
14. false
15. true
16. true
17. true
18. true
19. false
20. false
21. a
22. b
23. a
24. c
25. b
26. b
27. b
28. b
29.-34. Any order; examples:
29. lever — hammer
30. inclined plane — ramp
31. wedge — axe
32. wheel and axle — doorknob
33. pulley — block and tackle
34. screw — wood screw
35. a. Examples:
tractor, truck
b. Examples:
bus, car, airplane
c. Examples:
bulldozer, cement truck
d. Example:
ocean liner
36. Any order:
a. heat
b. light
c. sound
d. electricity
37. Either order:
a. stored energy
b. energy in action
38. a. simple
b. complex
39. Work is pushing or pulling something to move it.

ALTERNATE LIFEPAC TEST

1. d
2. g
3. b
4. h
5. a
6. j
7. i
8. e
9. c
10. f
11. true
12. true
13. false
14. true
15. false
16. true
17. true
18. false
19. b
20. c
21. a
22. b
23. c
24. a
25. b
26. b

27.-32. Examples; any order:

27. a. lever
b. hammer

28. a. inclined plane
b. ramp

29. a. wedge
b. axe

30. a. wheel and axle
b. wheelbarrow

31. a. pulley
b. block and tackle

32. a. screw
b. wood screw

33. Any order:
a. heat
b. light
c. sound
d. electricity

34. Either order:
a. stored energy
b. energy in action

35. a. simple
b. complex

36. Examples:
a. bus, auto, airplane
b. wheelbarrow, food cart
c. crane, elevator
d. steam shovel, oil drill

SCIENCE 404

ALTERNATE LIFEPAC TEST

NAME ______________________

DATE ______________________

SCORE ______________________

Match these items (each answer, 2 points).

1. _______ friction
2. _______ energy in action
3. _______ machine
4. _______ mechanical advantage
5. _______ mesh
6. _______ pitch
7. _______ slope
8. _______ load
9. _______ work
10. _______ stored energy

a. to interlock
b. tool to make work easier
c. moving an object
d. stops moving things
e. thing to be moved
f. energy not being used
g. energy in motion
h. force gained by using a machine
i. slant
j. distance between threads of a screw

Write *true* or *false* (each answer, 2 points).

11. ____________ By using a lever, a heavy load can be raised with less force.
12. ____________ A boy swimming is energy in action.
13. ____________ A wheel is a simple machine.
14. ____________ The use of gears increases speed.
15. ____________ A doorknob is an inclined plane.
16. ____________ A spiral stairway is an inclined plane.
17. ____________ All lifting work is done against the force called gravity.
18. ____________ The simple machine called a wedge is used only by woodcutters.

Write the letter of the correct answer in each blank (each answer, 3 points).

19. Heat, light, and sound are kinds of ____________________ .
a. friction b. energy c. gravity

20. For work to be done, you must overcome ____________________ .
a. gravity b. friction c. both a and b

21. A wheelbarrow is ____________________ .
a. a lever b. an inclined plane c. a wedge

22. A windlass is a ____________________ .
a. lever b. wheel and axle c. neither a nor b

23. You use gears in ____________________ .
a. bicycles b. eggbeaters c. both a and b

24. The inclined plane on a screw is called the ____________________ .
a. thread b. pitch c. direction

25. If pulleys are arranged with four strands of rope, the mechanical advantage is ____________________ .
a. 2 b. 4 c. 8

26. A scientist who learned to use steam energy was ____________________ .
a. Galileo b. Watts c. both a and b

Complete the following lists (each numbered item, 4 points).

	The six simple machines:	An example:
27.	a. ____________________	b. ____________________
28.	a. ____________________	b. ____________________
29.	a. ____________________	b. ____________________
30.	a. ____________________	b. ____________________
31.	a. ____________________	b. ____________________
32.	a. ____________________	b. ____________________

33. Name four kinds of energy.
a. ____________________
b. ____________________
c. ____________________
d. ____________________

34. Name two forms of energy.

a. ______________________________

b. ______________________________

35. Name the two kinds of machines.

a. S ______________________________

b. C ______________________________

36. Name a complex machine that might be used

a. for traveling.______________________

b. for pushing. ______________________

c. for lifting.________________________

d. for digging. ______________________

SCIENCE 405

Unit 5: Electricity and Magnetism

TEACHING NOTES

MATERIALS NEEDED FOR LIFEPAC	
Required	**Suggested**
• some iron fillings • a bar magnet • a large iron nail • a piece of steel (a knife blade or scissors) • strip of copper or copper wire • strip of zinc metal or galvanized tin • one lemon • some bell wire (light gauge solid copper wire) • a dry cell • a knife switch • small metal objects • string • two balloons • a woolen sweater or piece of woolen cloth • a rubber comb or hard rubber rod • one magnet • two paper clips • two rubber bands • five pins • one piece of chalk • two bits of wood • one small piece of paper • one penny • one nickel • one tin can • one small plastic cup • a compass • meter stick • about 40 centimeters of heavy string • a stack of books • a galvanometer • a mailing tube or the cardboard from a roll of paper towels	• 4th Grade Science Experiments video

ADDITIONAL LEARNING ACTIVITIES

Section 1: Electricity

1. Open and have students examine a dry cell.
2. Show students how to splice wires. Emphasize that insulation must be well scraped off, bare wires must be twisted tightly, and at least two layers of insulating tape wound around the splice. Have one or two students follow the demonstration with one of his own. Have class evaluate student's demonstration.
3. Have students compile a list of safety precautions beyond those listed in LIFEPAC. Have them consult parents, electricians, custodians, radio/TV repairmen, and other adults.
4. Have student demonstrate static electricity. Rub rubber rod with wool. Stick rod into box of puffed rice cereal. Student will notice that the puffed rice is attracted, then repelled. Discuss reaction of puffed rice.
5. Look up information and write about Heinrich Geissler (1814-1879) and Joseph Thompson (1856-1940). Both of these men found that charged particles glow and searched for the reason.
6. Have student find out what a Leyden Jar is and how static electricity can be stored in it.

Section 2: Magnetism

1. Have students prepare a table display and bulletin-board background on magnetism. Display experiments and appropriate books.
2. Make a list of all the ways you can think of that magnets are used. For instance, large magnets hung from boats pick up iron parts from sunken ships.
3. Tape a powerful horseshoe magnet to a piece of cardboard. Tape it about one-half of the way down from the top and about one inch from the center line. Then turn the cardboard over and prop it against a pile of books. Slide iron washers (or slugs) and pennies down the center line. Soft drink, candy, and other vending machines use a similar device to detect slugs and other kinds of fake money. Discuss cheating and dishonesty in relation to little things.
4. Bring toys and games from home that show how magnets are used in industry.

ANSWER KEYS

SECTION 1

1.1 Teacher check
1.2 false
1.3 true
1.4 true
1.5 false
1.6 true
1.7 Electrons are either removed from or added to an object.
1.8 neutral
1.9 They move away from each other
1.10 They move together or are attracted to each other.
1.11 c
1.12 b
1.13 e
1.14 a
1.15 d
1.16 f
1.17 They fly apart.
1.18 Like charges push each other away.
1.19 The water bends toward the charged object.
1.20 Either order;
a. electrolyte
b. electrodes
1.21 Teacher check
1.22 lemon
1.23 a. copper
b. zinc
1.24 Teacher check
1.25 no
1.26 Both metals are the same.
1.27 no
1.28 Both metals are the same.
1.29 Conductors
metal spoon, screwdriver handle and screwdriver blade, if they are made of metal, tin can, paper clip, lemon juice, vinegar, pencil, lead, pure water, salt water
Nonconductors
chalkboard eraser, glass dish, rubber ball, paper, wood, silk, wool, chalk
1.30 a material through which electricity will flow
1.31 a material through which electricity will not flow
1.32 Teacher check
1.33 a "track" for current which begins and ends at the same place
1.34 a switch
1.35 Example:
Would not have automobiles or electric ally cooled trucks to bring the food. Horses would have to draw wagons bringing food.
1.36 Would have to use a coal and/or wood stove.
1.37 Example:
I could use an ice box. Cellars into the ground used to be used to store food
1.38 Either order:
a. fuse
b. circuit breaker
1.39 The fuse is a thin piece of metal. The current runs through the metal. If there is too much current, the piece of metal will melt.
1.40 The cords could be damaged by heavy furniture and cause a fire.
1.41 un; to remove cover
1.42 re; to cover again
1.43 un; to remove load
1.44 re; to load again
1.45 re; to wind again
1.46 return
1.47 incorrect
1.48 unsafe
1.49 uncertain
1.50 unload
1.51 uncover

SELF TEST 1

1.01 d
1.02 j
1.03 a
1.04 f
1.05 k
1.06 g
1.07 c
1.08 e
1.09 b
1.010 h
1.011 b
1.012 a
1.013 c
1.014 a
1.015 b
1.016 c
1.017 b
1.018 a
1.019 b
1.020 false
1.021 true
1.022 true
1.023 true
1.024 true
1.025 true
1.026 false
1.027 false
1.028 true
1.029 The fuse is a container with a thin piece of metal. The current runs through the metal. If there is too much current, the piece of metal will melt.
1.030 Example:
I would probably eat fresh food. I would not have frozen popsicles. Soft drinks could not be chilled. Perishable food such as meat would have to be eaten immediately.
1.031 Worn or broken electrical cords should be replaced.
1.032 un
1.033 re
1.034 in
1.035 ex
1.036 in

SECTION 2

2.1 themselves
2.2 a. iron
b. steel
2.3 iron
2.4 brass
2.5 objects of iron and steel are attracted
2.6 no
2.7 yes
2.8 it picks up the filings
2.9 around the poles
2.10 Teacher check
2.11 a. attract iron and steel materials
b. point north
2.12 it should
2.13 it picks them up
2.14 The object dropped off the nail. The nail is a magnet only when it is surrounded by an electric current.
2.15 urge
2.16 score
2.17 north
2.18 chart
2.19 storm
2.20 worst
2.21 heard
2.22 Your dad wired your electric train set, and now you can play with it.
2.23 The circuit was broken so Mother called the repairman.
2.24 Because the object has more electrons than protons, the object has a negative electrical charge.
2.25 Because of the thunderstorm, the lights in the neighborhood went out.
2.26 Our electric lights are on because the man fixed the wiring in our new house.

SELF TEST 2

2.01 k
2.02 a
2.03 f
2.04 e
2.05 g
2.06 c
2.07 d
2.08 i
2.09 j
2.010 b
2.011 at the poles
2.012 iron and steel
2.013 compass
2.014 electrons
2.015 current electricity
2.016 true
2.017 true
2.018 true
2.019 true
2.020 false
2.021 false
2.022 true
2.023 false
2.024 false
2.025 true
2.026 c
2.027 b
2.028 c
2.029 a
2.030 c
2.031 Magnets can be made electric.
2.032 Electricity will flow through a conductor. Electricity will not flow through an insulator.
2.033 The fuse is a container with a thin piece of metal. The current seeps through the metal. If there is too much current, the piece of metal will melt.

LIFEPAC TEST

1. f
2. c
3. i
4. a
5. d
6. k
7. b
8. e
9. h
10. g
11. true
12. true
13. false
14. true
15. false
16. true
17. true
18. true
19. true
20. false
21. false
22. true
23. true
24. false
25. false
26. true
27. false
28. b
29. c
30. b
31. c
32. b
33. c
34. c
35. a. A conductor will carry electric current.
    b. Water and copper are two good conductors.
36. a. An insulator will not carry electricity.
    b. Rubber and glass are two insulators
37. a fuse is a safeguard to prevent an overload of electricity.

ALTERNATE LIFEPAC TEST

1. g
2. d
3. j
4. c
5. a
6. i
7. f
8. b
9. e
10. h
11. true
12. false
13. true
14. true
15. false
16. false
17. true
18. false
19. true
20. true
21. true
22. c. electrons
23. b. water
24. a. compass
25. c. Benjamin Franklin
26. b. made electric
27. a. circuit breaker
28. c. north
29. a. an insulator
30. c. poles
31. c. lightning
32. Examples:
    a. copper
    b. aluminum
33. generator
34. Example:
    A fuse prevents an overload of electricity and shuts off the current.
35. Unplug them and replace them.
36. It could catch fire.
37. the Bible
38. an atom

SCIENCE 405

ALTERNATE LIFEPAC TEST

NAME ______________________________

DATE ______________________________

SCORE ______________________________

Match these items (each answer, 2 points).

1. _______ electromagnet

2. _______ circuit

3. _______ poles

4. _______ fuse

5. _______ electron

6. _______ generator

7. _______ conductor

8. _______ insulator

9. _______ compass

10. _______ circuit breaker

a. particle of an atom

b. will not carry electricity

c. a safeguard

d. track for moving current

e. points north

f. will carry electricity

g. a magnet which is electric

h. switch

i. machine to make electricity

j. ends of a magnet

Answer *true* or *false* (each answer, 2 points).

11. ____________ Electricity is very important to our world.

12. ____________ Electricity will flow through an insulator.

13. ____________ Electricity is related to magnetism.

14. ____________ Electricity can be dangerous.

15. ____________ Charges that are different push each other away.

16. ____________ Protons always move out from the center of an atom.

17. ____________ A rubber comb is a poor conductor of electricity.

18. ____________ Electric current will not flow through water.

19. ____________ The greatest pull of a magnet is near its poles.

20. ____________ Electrons can move from atom to atom.

21. ____________ Lightning is mentioned in the Bible.

Write the letter for the correct answer on each line (each answer, 3 points).

22. When an object is electrically charged, ______________ have been added or removed from the object.
a. neutrons b. protons c. electrons

23. A good conductor of electricity is ______________ .
a. rubber b. water c. glass

24. When you go hiking in the woods, you should take a ______________ .
a. a compass b. tent c. dog

25. A man who experimented with electricity was ______________ .
a. John Smith b. Robert Jones c. Benjamin Franklin

26. *Charged* means ______________ .
a. made over b. made electric c. made empty

27. A switch is a ______________ .
a. circuit breaker b. fuse c. insulator

28. A natural compass always points ______________ .
a. south b. west c. north

29. Electricity will not travel through ______________ .
a. an insulator b. copper c. water

30. The ends of a magnet are called its ______________ .
a. spikes b. current c. poles

31. One form of electricity is ______________ .
a. clouds b. thunder c. lightning

Answer these questions (each problem, 4 points).

32. What are two conductors of electricity?

a. ______________________ b. ______________________

33. What is a machine called that makes electricity in large amounts?

____

34. How does a fuse help prevent fires?

____

____

____

35. What should you do with worn-out electrical cords on lamps and appliances?

____

____

36. What could happen to your house someday if it had no fuses?

____

____

37. What book has a chapter called *Job*?

____

38. What contains a particle called an electron?

____

SCIENCE 406

Unit 6: Properties of Matter

TEACHING NOTES

MATERIALS NEEDED FOR LIFEPAC	
Required	Suggested
• a plastic bowl large enough to hold 3 or 4 cups of water and 2 or 3 ice cubes • one small paper cup and enough crushed ice to fill the cup • one stalk of celery • a heavy-duty plastic knife • 1 clear plastic glass (disposable type) • red food coloring • 4 plastic glasses (1 for each test material) • 1 teaspoon of salt • 1 teaspoon of sugar • 1 teaspoon of salad or cooking oil • a plastic spoon or wooden stick for stirring • a small pan that holds 1 or 2 cups of water • 1/3 cup of water • 1 cup of sugar or salt • a piece of string • a paper clip or a small nail • 1 clear glass (must be round) • 1 clear freezer carton (must be square) • 1 paper bag (lunch bag is best) • 1 bottle of strong smelling perfume • 1 bottle of household ammonia (just a small amount is enough)	• additional library books about water and matter • 4th Grade Science Experiments video

ADDITIONAL LEARNING ACTIVITIES

Section 1: Water

1. Request from your local public library a teacher's collection of age-appropriate books about water and matter and have these available in the classroom for student use.
2. Visit a museum to learn about the development of the steam engine. One group reports on this topic, another on development of hydroelectric power. Students should use reference books to prepare questions they wish to explore. Each group reports findings to class.
3. Students trace chain references to forms of water in Scripture. One student takes topic on hail, one snow, and another on ice.
4. Students may prepare a colorful bulletin board that displays the fruits and vegetables that have large amounts of water.
5. Demonstrate to the class that it is possible to fill a glass with water above the top of the glass.

   a. Slowly fill the glass with water.
   b. Carefully continue to slowly add water until it fills the glass to the top and bulges a little above the rim.
   c. Drop straight pins one by one into the glass until it overflows into the saucer. (Shows result of surface tension: molecules attracting each other and sticking together.)

6. Student may give report on the use of steam in the development of steam engines.
7. Student may wish to report on how to make homemade ice cream. Ask him to explain why salt and ice are used to help the cream freeze.
8. Student may record the total amount of liquid he drinks in a day. From this figure, calculate how much he would drink for a week, a month, and a year.
9. An interested student may wish to make a survey of pollution in a nearby water supply and report on it.
10. As a class, do some brainstorming about the uses of water. List on the board as many uses as you can think of.
11. At home, have the students collect pictures from magazines or other sources about the uses of water. Combine as many as possible on a class collage bulletin board and label it "The many uses of water" or some other appropriate title.
12. Have each student read one outside book or article about water and its uses and give a brief oral report to the class or family. Allow students to question each other about their reports.
13. Together as a class, read aloud the biographies of Gabriel Fahrenheit and Anders Celsius and discuss how each man developed his thermometer scale.
14. If possible, tour a hydroelectric power plant and observe first hand how water is used in this type of facility.
15. Visit a water treatment plant to see how water is treated for home use and why this is necessary. Follow up any field trips with oral or written quizzes, class discussion, or short written reports, as you feel is needed to ensure understanding.
16. Discuss with the class various ways to conserve water. Encourage each student to develop a "home water conservation plan."
17. Discuss and give examples of the three states of water: Examples might include:
    liquid: rain, water from a faucet, a river
    solid: ice cubes, glaciers, icebergs, ice skating rinks, snow, or frost
    gas: water vapor or steam
18. Discuss and give examples of the three states of all types of matter: Examples are given as follows. Encourage the students to really think about it and brainstorm as many examples as possible to emphasize the point that matter is everywhere!
    liquid: juice, milk, gasoline, syrup
    solid: chair, car, building, glass
    gas: oxygen, hydrogen, carbon dioxide
    (Note: Students may need help at this age level in identifying types of gases. A basic discussion on gases could be used as a springboard to introduce the concept of elements and compounds that is covered in Section II of this LIFEPAC.)

19. Present and discuss the water cycle and the forms that water takes (solid, liquid, and gaseous) as it travels from the earth to the sky, then back to the earth again. (Note: The water cycle is presented in section 1 of Science LIFEPAC 407. A preliminary discussion at this point could introduce the concept and serve as a link for what is coming up in the next LIFEPAC.)
20. Discuss this "stumper" question: Does hot water freeze faster than cold water? (Yes!) Who discovered this? (This fact was discovered in 1969 by a Tanzanian [African] high school student named Mpemba.) What is this phenomena called? (The Mpemba effect.)

Section 2: Matter

1. Have pupils visit a museum gem collection. Visit the department that displays crystals.
2. Students may look up the history of chemistry and prepare a report on alchemists.
3. Student may make a circular graph showing the different kinds of gases in the air.
4. Examine some crystals under the microscope or with a magnifying glass. Salt crystals are cube shaped. Aluminum crystals are eight-sided.
5. Read good biographies about Robert Boyle and/or John Berzelius aloud or as assigned reading. Discuss together as a class their scientific findings in more depth.
6. Design a "Building Blocks of Matter" poster or bulletin board showing:
   atoms➔molecules➔elements➔compounds
   Include labeled pictures of common compounds such as water (H_2O) and salt (NaCl). Look up the chemical names of other common everyday elements and compounds and include these labeled pictures as part of the project.
7. If possible, invite a chemist to speak to the class about what he/she does and why this work is important.

ANSWER KEYS

SECTION 1

1.1 yes
1.2 Suggested answer: Ice floats on top of the water. Fish and other animals are able to escape to the bottom of lakes and ponds.
1.3 no
1.4 Example: Ice expands when it freezes. The ice filled more of the cup than the melted water, which did not expand.
1.5 Example: Ice floats because it is much lighter than liquid water. It is lighter because it expands as it freezes.
1.6 Teacher check
1.7 8/10 or 80 percent; Teacher check
1.8 "For as the rain cometh down, and the snow from heaven, and returneth not thither, but watereth the earth, and maketh it bring forth and bud, that it may give seed to the sower, and bread to the eater."
1.9 Either order:
  a. Rain and snow water the earth
  b. Rain and snow help provide the seed or bread
1.10 yes
1.11 The water went up small tubes.
1.12 seven
1.13 vapor
1.14 Either order:
  a. rain
  b. melting snow
1.15 evaporate
1.16 a - c Any order:
  a. rain
  b. snow
  c. sleet
  d. hail
1.17 food
1.18 Example:
  A person could get water from foods such as milk, watermelon, or other fruits and vegetables with high water content.
1.19 pieces, or particles, dust
1.20 Any order:
  a. Water breaks down food so it can be used by the body
  b. Blood, which is about 50% water, carries food to all parts of the body.
  c. Water carries off body wastes.
1.21 Suggested answer: I can know it is old because the Bible story in Genesis tells how God made water on the second day.
1.22 The water droplets came from water vapor in the air.
1.23 The water vapor in the air cooled on the can and changed back to liquid form.
1.24 true
1.25 true
1.26 true
1.27 false
1.28 false
1.29 true
1.30 false
1.31 false
1.32 redo, retrace
1.33 unaware, unplug
1.34 misbehave, miscount
1.35 prefix, preset
1.36 refasten
1.37 unfasten
1.38 rebuild
1.39 prepay
1.40 Cooled moisture in the air will condense.
1.41 When water freezes, it expands.
1.42 Ice floats because it expands.
1.43 Ice melts when it is heated.
1.44 Answers will vary. Example: When water vapor cools, liquid water is made.
1.45 yes
1.46 yes
1.47 yes
1.48 yes
1.49 no
1.50 no
1.51 no
1.52 c
1.53 a
1.54 e
1.55 d
1.56 b
1.57 g
1.58 f
1.59 Example: Water breaks down materials to be used by the body. It also carries dissolved food to different parts of the body through the blood, which is mostly water.
1.60 Teacher check
1.61 Alum is used for making pickles, for medicine, and for water treatment.

1.62 "And the Spirit and the bride say, Come. And let him that heareth say, Come. And let him that is athirst come. And whosoever will, let him take the water of life freely."
1.63 Teacher check

SELF TEST 1

1.01 iceberg
1.02 glacier
1.03 0° Celsius
1.04 Snow
1.05 Ice
1.06 Dew
1.07 Blood
1.08 suspension
1.09 100° Celsius
1.010 expands
1.011 a
1.012 c
1.013 a
1.014 b
1.015 a
1.016 b
1.017 c
1.018 c
1.019 b
1.020 c
1.021 true
1.022 true
1.023 false
1.024 false
1.025 false
1.026 true
1.027 false
1.028 true
1.029 false
1.030 true

SECTION 2

2.1 b
2.2 c
2.3 b
2.4 c
2.5 a
2.6 Friend check
2.7 round
2.8 square
2.9 It is the shape of the bag.
2.10 Any order: solid, liquid, gas
2.11 container
2.12 gas
2.13 round
2.14 square
2.15 b. Matter takes up space.
e. All matter has mass or weight.
2.16 Is it a Solid, Liquid, or Gas?
liquid
solid
gas
Shape It Takes
container
definite
container
2.17 true
2.18 true
2.19 false
2.20 true
2.21 false
2.22 true
2.23 Teacher check
2.24

O	A	C	W	C	H	A	L	Z	O
X	C	B	C	L	E	F	K	Y	C
Y	Q	R	V	S	L	L	M	D	T
G	B	Z	C	N	I	Z	N	G	R
E	V	C	L	R	U	I	R	O	N
N	O	G	R	A	M	N	S	L	T
I	O	D	I	N	E	C	F	D	A
N	E	G	O	R	D	Y	H	T	U
C	Q	L	H	S	I	L	V	E	R
B	V	C	I	W	Y	G	S	R	L
N	C	T	F	C	O	B	A	L	T
T	I	N	V	W	R	U	S	U	V

2.25 elements
2.26 100
2.27 silver
2.28 penny
2.29 atom
2.30 H
2.31 O
2.32 compound
2.33 Teacher check
2.34 a. guitar
b. scarlet
c. star
d. large
e. farm
f. hard
g. third
h. skirt
i. girl
j. circle
k. stir
l. shirt
2.35 er
2.36 ur
2.37 or
2.38 ar
2.39 eer
2.40 er
2.41 eer
2.42 or
2.43 ur
2.44 ar
2.45 no or not
2.46 bad or wrong
2.47 backward or again
2.48 before
2.49 mismatch
2.50 relearn
2.51 preview
2.52 unwrap
2.53 unhappy
2.54 repack

SELF TEST 2

2.01 f
2.02 a
2.03 c
2.04 b
2.05 j
2.06 k
2.07 e
2.08 d
2.09 i
2.010 h
2.011 g
2.012 a
2.013 c
2.014 a
2.015 b
2.016 a
2.017 b
2.018 c
2.019 b
2.020 b
2.021 c
2.022 true
2.023 true
2.024 false
2.025 true
2.026 true
2.027 true
2.028 false
2.029 true
2.030 false
2.031 false
2.032 false
2.033 true

LIFEPAC TEST

1. c
2. a
3. j
4. d
5. e
6. k
7. f
8. h
9. b
10. i
11. false
12. true
13. false
14. true
15. true
16. true
17. false
18. false
19. true
20. true
21. true
22. true
23. gas
24. properties
25. 0°
26. compound
27. matter
28. solvent
29. Blood
30. definite
31. 212°
32. Clouds
33. water

ALTERNATE LIFEPAC TEST

1. f
2. j
3. h
4. a
5. k
6. e
7. b
8. g
9. c
10. i
11. true
12. true
13. true
14. true
15. false
16. true
17. false
18. true
19. false
20. true
21. true
22. false
23. insoluble
24. suspension
25. 100°
26. Atoms
27. penny
28. molecule
29. Dew
30. shape
31. 32°
32. Clouds
33. nutrients

SCIENCE 406

ALTERNATE LIFEPAC TEST

NAME ____________________

DATE ____________________

SCORE ____________________

Match these items (each answer, 3 points).

1. _______ building blocks of matter which make molecules
2. _______ the science that deals with earth materials
3. _______ to take up more space
4. _______ a large amount of ice moving on land
5. _______ a floating mountain of ice
6. _______ a very little bit
7. _______ a material that can dissolve other materials
8. _______ a cleaning liquid
9. _______ a material made up of two or more elements
10. _______ made of only one kind of atom

a. glacier
b. solvent
c. compound
d. dew
e. particle
f. atoms
g. wood alcohol
h. expand
i. element
j. chemistry
k. iceberg

Answer *true* or *false* (each answer, 3 points).

11. ____________ Weight, size, and color are all properties of matter.
12. ____________ Heated water becomes steam in a gas form.
13. ____________ Matter can be a solid, liquid, or gas.
14. ____________ The human body is 70% water.
15. ____________ People are not made up of matter.
16. ____________ Steam is a gas form of water.
17. ____________ A compound is the smallest particle of a material that is still like that material.
18. ____________ Dew is condensed water vapor.

19. __________ Molecules cannot be split into smaller pieces.
20. __________ Heating a solution allows it to dissolve more material.
21. __________ Atoms are joined together to make molecules.
22. __________ Oil is soluble in water.

Fill in the blanks using words from the Word Bank (each answer, 3 points).

suspension	clouds	penny	molecule
nutrients	insoluble	100°	dew
atoms	32°	gas	shape

23. A material that does not dissolve in another is known as a(n) __________ material.
24. Sand mixed with water creates a(n) __________ .
25. Water boils at __________ Celsius.
26. __________ are joined together to make molecules.
27. A(n) __________ contains the element copper.
28. A(n) __________ is made up of atoms joined together.
29. __________ is cooled water vapor.
30. Matter in liquid or gas form takes the __________ of the container it is in.
31. Water freezes at __________ Fahrenheit.
32. __________ are made of water vapor.
33. Water carries __________ to all parts of the bodies of people and animals.

SCIENCE 407

Unit 7: Weather

TEACHING NOTES

MATERIALS NEEDED FOR LIFEPAC	
Required	Suggested
• a water glass • a pan full of water • drinking glass full of water • piece of cardboard large enough to cover the glass • sink or large basin to catch the water after the experiment • sheet of black paper • window • sunlight shining through the window	• 4th Grade Science Experiments video

ADDITIONAL LEARNING ACTIVITIES

Section 1: Cause of Weather

1. After a rain ask students what happens on the grass and sidewalk. Have them observe evaporation by placing a little water in a low dish in the sunlight.
2. Make a class scrapbook of poems about the weather. Try to find at least one poem for each month. Students gifted in art may illustrate the book. Book may be donated to a children's hospital or to an ill student accompanied by letter(s).
3. Make bulletin board on causes of weather.
4. Find out how air pressure affects weather. What is a front? What kind of weather does a high-pressure front bring? A low-pressure front?
5. After observing clouds on a particular day, make imaginary cloud pictures. Write an adventure story to go with the pictures.
6. Keep a record of the weather for two weeks. Make a chart for the classroom.

Section 2: Forces of Weather

1. Hold discussion of "How Weather Helps Us."
   a. How can rainy weather be helpful to plants?
   b. How does rain help people?
   c. What does lightning do to the soil that is helpful?
   d. What kinds of recreation does the wind provide?
   e. What kinds of fun can you have in the snow?
2. Several students may enjoy making a display for a book corner entitled "Rainy Days Are Good Reading Days." Include some good library books.
3. Paint a picture of different kinds of weather. Make a classbook of the pictures. Give it a title.

4. Collect pictures from old magazines of different kinds of weather. Organize into sections such as, rainstorms, hurricanes, tornadoes, snowstorms, or lightning.
5. Find out how to make six-pointed snowflakes from paper. Cut out and hang in classroom. Why are all snowflakes six-pointed?
6. Write a story about "A Raindrop" or "A Snowflake."

Section 3: Prediction of Weather

1. Visit a weather station at the local airport to find an answer to questions related to weather prediction.
2. Play tape recording previously recorded at home of a weather report. Discuss with class.
3. Some students may be interested in making a weather vane. Use library book or encyclopedia for directions. Tell why it is important to know from what direction the wind if blowing.
4. Have students explain the difference between liquid and aneroid barometers. Show each kind in class if available.
5. Student can give oral report of Gabriel Fahrenheit.
6. Student watch a weather report on television and tell about it in class.
7. Find pictures showing relationship between weather and geography and show in class.

ANSWER KEYS

SECTION 1

1.1 atmosphere
1.2 troposphere
1.3 ozone
1.4 sun
1.5 Choose from: troposphere, exosphere, ionosphere, stratosphere, or mesosphere
1.6 a. across the land and water
b. up and down
1.7 Example:
It thins out (gets thinner and thinner).
1.8 Examples:
a. balloons
b. airplanes
c. sailboats or gliders, kites
1.9 second day
1.10 Hint: The air in the glass pushed some of the water down.
1.11 It returned to the same level as it was in the beginning.
1.12 no
1.13 because of the air in the glass
1.14 air fills space. (The air takes the place of water.)
1.15 yes
1.16 no
1.17 The window glass is transparent. The paper is not.
1.18 radiation
1.19 air (atmosphere)
1.20 transparent
1.21 transparent
1.22 nearest (closest)
1.23 cooler
1.24 three and one-half
1.25 two
1.26 Heat rays from the sun pass through space.
1.27 Heat rays from the sun pass through our atmosphere.
1.28 Heat rays strike the ground and warm it.
1.29 The ground absorbs heat from the sun.
1.30 The ground heats the air above it.
1.31 no
1.32 no
1.33 It must have been the air (pressure) that is pushing against the card board. Air pushes from all directions.
1.34 c. moon
1.35 c. equal to
1.36 b. air
1.37 a. altitude
1.38 false
1.39 true
1.40 false
1.41 true
1.42 true
1.43 true
1.44 false
1.45 false
1.46 moisture
1.47 fog
1.48 steam
1.49 cloud
1.50 vapor
1.51 d
1.52 b
1.53 f
1.54 c
1.55 a

SELF TEST 1

1.01 g
1.02 f
1.03 i
1.04 a
1.05 e
1.06 d
1.07 b
1.08 c
1.09 rises
1.010 radiation
1.011 cycle
1.012 fog
1.013 troposphere
1.014 expands
1.015 pressure
1.016 second
1.017 atmosphere
1.018 c. air
1.019 c. water
1.020 b. steam
1.021 a. vapor
1.022 c. ground
1.023 a. rises
1.024 b. ultraviolet
1.025 a. troposphere
1.026 b. altitude
1.027 true
1.028 false
1.029 true
1.030 true
1.031 false
1.032 false
1.033 true
1.034 false
1.035 true
1.036 Any order:
a. moisture in the air
b. temperature
c. wind changes
d. air pressure changes

SECTION 2

2.1 true
2.2 true
2.3 true
2.4 false
2.5 false
2.6 true
2.7 false
2.8 c
2.9 d
2.10 a
2.11 f
2.12 b
2.13 c
2.14 d
2.15 c
2.16 a. God Gave Us a Beautiful Gift When He Gave Us Snow.
b. God created snow for us to enjoy.
c. Snow's whiteness reminds us of a Bible verse
d. Isaiah told us that our sins could become white like snow.
e. Children Who Live Where Snow Does Not Fall Are Missing Something.
f. They cannot go sledding.
g. They cannot go skiing.
h. They cannot make a snowman.
i. You Can Learn Much about Animals from Tracks in the Snow.
j. You can tell what animals they were.
k. You may be able to tell how fast they were going.
l. Snowflakes Have Beautiful Designs.
m. They are star-like.
n. They have six points.
o. No two are alike.
2.17 Teacher check
2.18 Examples: Sand and dust storms can be dangerous to drivers because they prevent them from seeing the road. An accident may result. Or Dust storms are sometimes harmful because they blow crops out of the ground and carry the soil miles away.
2.19 a. gone down
b. upward
c. hungry
d. windy; stormy
2.20 Teacher check
2.21 surface
2.22 erosion
2.23 sand
2.24 geographic

SELF TEST 2

2.01 i
2.02 e
2.03 j
2.04 b
2.05 h
2.06 f
2.07 a
2.08 c
2.09 d
2.010 ozone
2.011 radiation
2.012 water
2.013 irrigation
2.014 waterspouts
2.015 dangerous
2.016 geography
2.017 fog
2.018 b. five
2.019 c. women and men
2.020 a. heats
2.021 c. atmosphere
2.022 b. fertilizer
2.023 c. richer
2.024 b. not safe
2.025 a. satellites
2.026 c. warm
2.027 true
2.028 true
2.029 false
2.030 true
2.031 true
2.032 false
2.033 false
2.034 true
2.035 false
2.036 true
2.037 true

SECTION 3

3.1 b. predicting
3.2 a. weather
3.3 Teacher check
3.4 Example: 30 degrees below zero
3.5 Example: 250 degrees
3.6 Teacher check
3.7 Example: 100 degrees
3.8 Example: 0 degrees
3.9 0 degrees C
3.10 c
3.11 g
3.12 a
3.13 b
3.14 d
3.15 e
3.16 ba rom' e ter
3.17 e vap' o rate
3.18 fer' ti li zer
3.19 ge o graph' ic
3.20 ra di a' tion
3.21 ther mom' e ter
3.22 four

SELF TEST 3

3.01 c
3.02 f
3.03 h
3.04 b
3.05 e
3.06 j
3.07 a
3.08 i
3.09 d
3.010 heat
3.011 troposphere
3.012 predict
3.013 weather
3.014 temperature
3.015 centigrade
3.016 b. boiling
3.017 b. weather
3.018 c. predicting
3.019 b. pressure
3.020 c. rays
3.021 a. ground
3.022 c. vane
3.023 c. barometer
3.024 a. centigrade
3.025 a. second
3.026 b. rocks
3.027 c. thermometer
3.028 true
3.029 false
3.030 true
3.031 false
3.032 true
3.033 true
3.034 false
3.035 People have been hurt in a lightning storm. Trees and buildings have been damaged. The storm is helpful because fertilizer for soil is formed as lightning causes gases to blend and fall to the ground as fertilizer.

LIFEPAC TEST

1. g
2. d
3. i
4. h
5. c
6. j
7. e
8. a
9. f
10. b
11. direction
12. Weather
13. predict
14. eye
15. any woman's or man's name
16. transparent
17. b. centigrade
18. a. troposphere
19. c. air pressure
20. a. zero
21. b. slower than
22. c. wind force and speed
23. c. soil
24. a. Heat rays from the sun pass through our atmosphere.
b. Heat rays strike the ground and warm it.
c. The ground absorbs heat from the sun.
d. The ground heats the air above it.
25. Examples: Changes in the atmosphere, temperature, air pressure, air movements, and moisture cause weather to change.
26. Rainstorms happen when a large amount of water gathers in a cloud and falls to earth. Rainstorms supply water but sometimes cause bad floods. Blizzards are heavy, wind-blown snowstorms that block roads. The helpful part is the supply of water from the melting snow and cover for plants.
27. false
28. true
29. true
30. false
31. true

ALTERNATE LIFEPAC TEST

1. e
2. d
3. j
4. i
5. h
6. a
7. f
8. g
9. b
10. c
11. vane
12. weather
13. satellite
14. sun
15. fog
16. snow
17. thermometer
18. water
19. electricity
20. 32
21. true
22. false
23. true
24. false
25. true
26. true
27. false
28. false
29. true
30. true
31. true
32. a. air
33. c. barometer
34. b. centigrade
35. b. sound
36. c. center
37. c. soil
38. a. erosion
39. c. warm
40. a. heat
41. c. space
42. b. rainfall
43. c. air
44. Either order:
a. air pressure changes
b. temperature changes or wind currents, wind, or moisture
45. Example: A hurricane blows down buildings, trees, and telephone poles. It sometimes brings floods and kills people and animals.

SCIENCE 407

ALTERNATE LIFEPAC TEST

NAME ______________________________

DATE ______________________________

SCORE ______________________________

Match each item (each answer, 2 points).

1. _______ weather vane
2. _______ tornado over water
3. _______ barometer
4. _______ weather bureau
5. _______ violent snow storm
6. _______ erosion
7. _______ richer soil
8. _______ air temperature
9. _______ ultraviolet rays
10. _______ rain gauge

a. geographic force
b. ozone
c. measures rainfall
d. waterspout
e. wind direction
f. lightning
g. thermometer
h. blizzard
i. predicts weather
j. measures air pressure

Complete these statements (each answer, 3 points).

11. The direction of the wind is shown by an instrument called a weather or wind _______________ .
12. Although we cannot control the _______________ , we can predict it better when we understand its forces.
13. The weather bureau takes pictures of clouds in the sky by means of a _______________ in space.
14. The ozone layer of gas around the earth protects us from the ultra-violet rays of the _______________ .

15. A cloud of water vapor on or close to the ground is called _______________ .

16. A blizzard is a storm with strong winds and large amounts of _____________ .

17. An instrument for measuring temperature is called a _______________ .

18. Steam is very hot _______________ .

19. It is dangerous to swim outdoors during a lightning storm because _______________ travels through water.

20. Water freezes at _______________ degrees Fahrenheit.

Write *true* or *false* (each answer, 1 point).

21. _____________ In the daytime the ground becomes warmer because of heat from the sun.

22. _____________ A waterspout is used to measure rain.

23. _____________ Thunderstorms help fertilize the earth and make the soil richer.

24. _____________ A weather vane tells the speed of the wind.

25. _____________ A barometer tells what the air pressure is.

26. _____________ Lightning heats the air it travels through.

27. _____________ Sound travels faster than light.

28. _____________ There is a small amount of air on the moon.

29. _____________ Warm air always rises.

30. _____________ A car is a safe place to be during a thunderstorm.

31. _____________ A tornado is made of air.

Write the letter for the correct answer on each line (each answer, 2 points).

32. The layer of _______________ which lies closest to the earth is called the troposphere.
a. air b. water c. vapor

33. Air pressure is sometimes measured by an aneroid _______________ .
a. thermometer b. wind gauge c. barometer

34. Water freezes at zero degrees _______________ .
a. Fahrenheit b. Centigrade

35. Light travels faster than _______________ .
a. lightning b. sound c. light

36. The "eye" of a hurricane is in the _______________ of it.
a. outside b. bottom c. center

37. Thunderstorms that water the earth keep ____________ fertile.
a. fertilizer b. water c. soil

38. Wind and water are geographic forces that cause ____________ .
a. erosion b. dryness c. oceans

39. Heat rays from the sun strike the ground and ____________ it.
a. cool b. wear c. warm

40. The ground absorbs ____________ from the sun.
a. heat b. light c. air

41. Heat rays pass through ____________ to reach the earth.
a. water b. land c. space

42. A rain gauge measures the amount of ____________ that takes place in a certain length of time.
a. air b. rainfall c. lightning

43. We live in an ocean of ____________ .
a. water b. liquid c. air

Answer these questions (each answer, 5 points).

44. What are two causes of weather?
a. ____
b. ____

45. Tell how a hurricane causes damage. ____
____
____
____
____

SCIENCE 408

Unit 8: Our Solar System and The Universe

TEACHING NOTES

MATERIALS NEEDED FOR LIFEPAC	
Required	Suggested
• flashlight • rubber ball (about 6 inches in diameter) • a large white sheet of paper • square box with cover • large flashlight • star maps to copy • several sheets of black construction paper • pins	• 4th Grade Science Experiments video

ADDITIONAL LEARNING ACTIVITIES

Section 1: A Trip Through Space

1. Discuss with the students some of the latest explorations into space. Determine what false concepts some have received from reading or listening to science fiction. Discuss the difference between opinions not based on fact in contrast to the Word of God. What part does faith play in our beliefs? Are there some things we must accept by faith? As future scientists, what should be our attitude toward God and the Bible?
2. Read Psalm 19 aloud as the students listen. Then have them read it together and discuss it in relation to the wonders of God and the work of Creation.
3. Show several pictures of different types of telescopes and explain how they differ from each other.
4. Several members of a group may wish to make up a game called "True or False." Collect true facts from books and make up false statements and write them on cards. Shuffle and have each student draw a card one by one from the pile. The one drawing the card is to tell whether he is reading a true statement or a false one (The correct answer should be written on a key which is held by the leader). The person getting the largest number of cards is the winner.
5. Using directions from a reference book, some students may wish to make a model of the solar system.
6. Student may write a short fiction story about a trip to one of the planets.
7. Find out how far it is to each planet from the earth and determine how long it would take to reach each of them traveling at the speed of light.
8. Read some myths about the origin of the earth.
9. Read a book about Galileo and write a short book report about it.
10. Give a special oral report about the radio telescope.

Section 2: The Sun and Planets

1. Discuss Genesis 1:14 and Psalm 8:3 and 4 with the students. Ask them to tell why they are glad God created such a wonderful world.
2. Read the entire Creation story from Genesis and discuss with the students.
3. Arrange a bulletin board display entitled "A Flight to the Moon." Display interesting library books under it.
4. Refer to an atlas and draw a picture of the surface of the moon. Discuss the lakes and mountains. One student may draw the moon, another identify the "lakes" and craters.
5. Each member of the group may wish to select one feature of the moon and tell about it.
6. Draw a picture of what you think the earth would look like if gravity were removed.
7. Look up information on several of the largest meteorite craters in the world and give a report to the class.
8. Look up in an encyclopedia or online about Halley's comet. Tell about it in class.
9. Read the Christmas story aloud from the Bible (Matthew 1:18-2:11) and discuss it with the class.
10. Children may wish to dramatize the Christmas story.
11. Work together and give a flannel-board demonstration of the Christmas story.
12. Write as many reasons as you can think of why boys and girls should accept Jesus as their Lord and Saviour. Share your reasons with someone else.

ANSWER KEYS

SECTION 1

1.1 a. Mercury
b. Venus
c. Earth
d. Mars
e. Jupiter
f. Saturn
g. Uranus
h. Neptune (Based on the average distance from the sun.)

1.2 false
1.3 true
1.4 true
1.5 true
1.6 false
1.7 d
1.8 e
1.9 b
1.10 a
1.11 c
1.12 second
1.13 two
1.14 120,000
1.15 eight
1.16 c. an elephant
1.17 a. fact
1.18 c. truth
1.19 d. God
1.20 c. truth
1.21 a. Hans Lippershey
1.22 b. Galileo
1.23 d. spectroscope
1.24 He must observe, record, compare, and understand.
1.25 Example:
God gave man a mind to use the information he receives. If man cannot understand what he sees and hears, then the information will do him no good.
1.26 Friend check

SELF TEST 1

1.01 f
1.02 d
1.03 h
1.04 i
1.05 a
1.06 j
1.07 e
1.08 b
1.09 g
1.010 c
1.011 23
1.012 Mercury
1.013 eight
1.014 Mercury
1.015 astronomer
1.016 elephant or turtle
1.017 spectroscope
1.018 telescope
1.019 radio telescope
1.020 truth
1.021 d. Mercury and Venus
1.022 b. 240,000
1.023 d. galaxy
1.024 a. twenty-three
1.025 b. God
1.026 c. lenses
1.027 c. rainbow
1.028 b. radio
1.029 d. mind
1.030 b. Uranus and Neptune
1.031 Any five; any order:
a. Mercury
b. Venus
c. Earth
d. Mars
e. Jupiter, Saturn, Uranus, Neptune

1.032 true
1.033 false
1.034 true
1.035 true
1.036 false
1.037 false
1.038 true
1.039 true
1.040 true
1.041 true
1.042 He must observe, record, compare, and understand.

1.043 Any two; any order:
craters, mountains, or plains

1.044 Example:
It is very hot. The sun has a surface temperature of 11,000 degrees Fahrenheit.

SECTION 2

2.1 solar
2.2 life
2.3 revolves
2.4 seasons
2.5 plants
2.6 freeze
2.7 God's
2.8 93
2.9 stars
2.10 animals
2.11 true
2.12 false
2.13 false
2.14 true
2.15 true
2.16 false
2.17 true
2.18 true
2.19 Example: The ancient people did not know the truth about the sun; therefore, they made up stories about the sun from what they had observed. They did not have instruments like the telescope to find out the truth about the sun.
2.20 "And God made two great lights; the greater light to rule the day, and the lesser light to rule the night: he made the stars also."
2.21 moon
2.22 sun
2.23 d
2.24 e
2.25 a
2.26 b
2.27 c
2.28 c
2.29 a
2.30 b
2.31 d
2.32 b
2.33 force
2.34 sun
2.35 greater
2.36 planets
2.37 Newton
2.38 light
2.39 false
2.40 true
2.41 false
2.42 false
2.43 true
2.44 true
2.45 b. gravity

2.46 d. scientist
2.47 a. light
2.48 Teacher check
2.49 false
2.50 false
2.51 true
2.52 false
2.53 true
2.54 true
2.55 false
2.56 d
2.57 e
2.58 a
2.59 c
2.60 b
2.61 Either order:
a. metal
b. stone
2.62 tail
2.63 falling stars
2.64 a grain of sand
2.65 Either order:
a. Mars
b. Jupiter
2.66 a. an / oth / er
b. dis /cov / er
c. di / vert / ing
d. gal / ax / y
e. grav / i / ty
f. his / tor / y
g. im / por / tant
h. in / clud / ing
i. mys / ter / y
j. re / a / lize
k. reg / u / late
l. sci /en / tist
m. un / der / stand
n. u / ni / verse
2.67 Teacher check
2.68 From "Earth" the whole face of the ball is lit.
2.69 full
2.70 none of the moon can be seen
2.71 new
2.72 one-quarter
2.73 earth
2.74 240,000
2.75 Either order:
a. air
b. water
2.76 one-sixth
2.77 astronauts
2.78 rocks
2.79 f
2.80 e
2.81 b
2.82 a
2.83 c
2.84 Any four, any order:
a. no air
b. no water
c. too cold at night
d. too hot in daytime or lack of good soil no plant life
2.85 20 pounds
2.86 390 pounds
2.87 Examples:
gravity - the force that causes objects to move toward each other
astronomy - study of the stars, planets, galaxies, and the universe exploration - traveling to little known places to discover
astronauts - men and women who go into space to explore
astronomer - person who studies the universe

SELF TEST 2

2.01 million
2.02 second
2.03 eight
2.04 240,000
2.05 six
2.06 1969
2.07 star
2.08 sun
2.09 Mercury
2.010 planets
2.011 c. universe
2.012 c. gravity
2.013 a. oval
2.014 b. scientist
2.015 a. spectroscope
2.016 c. gravity
2.017 a. greater than
2.018 b. 865,000
2.019 c. God
2.020 a. earth
2.021 c
2.022 j
2.023 a
2.024 h
2.025 e
2.026 d
2.027 i
2.028 b
2.029 k
2.030 g
2.031 true
2.032 false
2.033 false
2.034 true
2.035 true
2.036 false
2.037 false
2.038 true
2.039 false
2.040 true
2.041 Either order:
a. spectroscope
b. radio telescope or telescope
2.042 Stars twinkle and planets have a steady light.
2.043 a. Mercury
b. Venus
c. Earth
d. Mars
e. Jupiter
f. Saturn
g. Uranus
h. Neptune
2.044 Any order:
a. no atmosphere
b. no food
c. no water or too cold, too hot

SECTION 3

3.1 e
3.2 d
3.3 b
3.4 f
3.5 g
3.6 a
3.7 c
3.8 c. 5,000
3.9 a. rotates
3.10 c. nine
3.11 b. North Star
3.12 a. Leo
3.13 c. Jesus
3.14 a. decorate
3.15 c. universe
3.16 b. Jesus
3.17 c. Big Dipper
3.18 a. Lord
3.19 b. before Christ
3.20 Example:
Astronomy is a true science; astrology is not.
3.21 Example:
People study astronomy to learn more about the sun, moon, stars, and the universe.
3.22 Example:
Scientists consider astrology to be untrue because it is based on falsehood. They have studied the stars and know that the stars cannot help people.
3.23 Example:
No one, especially Christians, should go to astrologists or study the signs of the zodiac to find out about the future. God has warned us not to believe in the false science of astrology. He has warned that nations would be ruined if they continued to worship false gods.
3.24 Across:
2. galaxy
5. infinite
7. constellations
8. astronomy

Down:
1. garnish
3. astrology
4. axis
6. firmament

3.25 c. stars
3.26 b. Milky Way
3.27 a. galaxies
3.28 a. very slowly
3.29 c. 250
3.30 b. stellar
3.31 b. mind
3.32 c. heaven
3.33 c. God
3.34 a. Jesus
3.35 false
3.36 false
3.37 true
3.38 false
3.39 true
3.40 false
3.41 true
3.42 true

SELF TEST 3

3.01 b. false
3.02 c. Dog Star
3.03 c. six
3.04 a. Mercury
3.05 a. 186,282
3.06 a. the North Star
3.07 b. twenty-three
3.08 c. earth
3.09 b. God
3.010 c. Star of the East
3.011 mind
3.012 Jesus (Christ)
3.013 Mars
3.014 spectroscope
3.015 away
3.016 the Milky Way
3.017 Neptune
3.018 go up
3.019 astronomy
3.020 minor planets
3.021 j
3.022 e
3.023 f
3.024 b
3.025 k
3.026 c
3.027 g
3.028 a
3.029 h
3.030 d
3.031 false
3.032 true
3.033 false
3.034 true
3.035 true
3.036 true
3.037 false
3.038 false
3.039 false
3.040 false
3.041 Example: A galaxy is a system, or group of stars in space.
3.042 Example: Heaven and earth will pass away.
3.043 Example: Astrology is a false science; astronomy is a true science. Astrology is superstition; astronomy is truth.
3.044 Either order:
a. supplies Vitamin D
b. makes plants grow

LIFEPAC TEST

1. c. false
2. b. gravity
3. c. God
4. b. Saturn
5. c. Star of the East
6. a. gravity
7. a. spectroscope
8. c. Mercury
9. a. away from
10. c. sun
11. constellations
12. telescope
13. astronomer
14. Mercury
15. Jupiter
16 axis
17. comet
18. radio telescope
19. sun
20. Mars
21. d
22. f
23. c
24. i
25. a
26. k
27. b
28. j
29. e
30. g
31. true
32. false
33. false
34. false
35. true
36. true
37. false
38. false
39. false
40. true
41. A fact is based on the truth. An opinion is a belief about something that might not be true.
42. Example:
Astronomy is a true science; astrology is a false science.
43. Any order:
a. weather
b. gravity
c. life
d. water or clouds, atmosphere
44. Either order:
a. Star of the East
b. North Star

ALTERNATE LIFEPAC TEST

1. b. moon
2. a. axis
3. b. telescope
4. a. galaxies
5. c. false
6. c. God
7. b. Newton
8. c. Mercury
9. a. six
10. a. telescope
11. c. spectroscope
12. b. gravity
13. a. planets
14. c. solar system
15. c. Christian
16. Jesus
17. God
18. star
19. Jupiter
20. Pluto
21. planets
22. comet
23. sun
24. e
25. g
26. b
27. a
28. j
29. l
30. k
31. f
32. h
33. c
34. i
35. d
36. true
37. false
38. false
39. false
40. true
41. false
42. false
43. true
44. false
45. a. Mercury
    b. Venus
    c. Earth
    d. Mars
    e. Jupiter
    f. Saturn
    g. Uranus
    h. Neptune
46. Examples; any order:
    a. clouds
    b. life
    c. water
    d. good soil

SCIENCE 408

ALTERNATE LIFEPAC TEST

NAME ________________________________

DATE ________________________________

SCORE ________________________________

Write the correct letter on the line (each answer, 2 points).

1. Astronauts who landed on the _______ brought back samples of rocks.
a. planet Mars b. moon c. earth

2. An imaginary line passing through an object and about which the object turns is called the _______ .
a. axis b. diameter c. coma

3. The instrument that has made it possible to map the stars is the _____ .
a. microphone b. telescope c. microscope

4. The _______ are moving away from each other at high speed.
a. galaxies b. planets c. asteroids

5. Astrology is a (n) _______ science.
a. up-to-date b. true c. false

6. The universe was created by _______ .
a. no one b. evolution c. God

7. Sir Isaac _______ was an astronomer who discovered the laws of gravity.
a. Chase b. Newton c. James

8. The planet nearest the sun is _______ .
a. Mars b. Earth c. Mercury

9. The planet, Saturn, has at least _______ rings.
a. six b. five c. ten

10. Lippershey invented the first _______ .
a. telescope b. airplane c. microscope

11. An instrument used by astronomers to measure light from distant objects is called a ______ .
a. telescope b. radio-telescope c. spectroscope

12. The natural force that causes objects to move toward each other is called _______.
a. astronomy   b. gravity   c. astrology

13. Astronomers are able to map the paths of the _______ because the universe is one of order.
a. planets   b. sea   c. winds

14. The sun is the center of the _______ .
a. galaxy   b. Milky Way   c. solar system

15. Changes on the earth or in the universe should not frighten _______ .
a. nonbelievers   b. sinners   c. Christians

Complete these statements (each answer, 3 points).

16. The Wise Men were guided to the baby _______________ by the Star of Bethlehem.

17. The universe was created by _________ on the fourth day.

18. The constellations were called _________ pictures and were imagined by the ancients.

19. The largest planet is ________________ .

20. The planet that was reclassified to a dwarf planet is _____________ .

21. Eight _______________ are revolving around the sun.

22. A heavenly body with a star-like center and a tail is called a ________________ .

23. The earth revolves around the _________ .

Match each item (each answer, 2 points).

24. _______ Lippershey
25. _______ Wise Men
26. _______ North Star
27. _______ Saturn
28. _______ God
29. _______ gravity
30. _______ radio-telescope
31. _______ minor planets
32. _______ elliptical
33. _______ opinion
34. _______ zodiac
35. _______ Egyptian god

a. at least six rings
b. pole star
c. belief
d. Ra
e. telescope
f. asteroids
g. Star of the East
h. oval
i. astrologers
j. created everything
k. radio waves
l. Newton
m. 93,000,000

Write *true* or *false* (each answer, 1 point).

36. ____________ The constellation that appears in the Northern Hemisphere in the months of April, May, and June is Leo.
37. ____________ The abbreviation for "before Christ" is A.D.
38. ____________ Astrology is a true science.
39. ____________ We cannot tell a planet from a star.
40. ____________ The Psalmist David praised the Lord for the Creation of the universe.
41. ____________ The earth is too far away from the sun for comfort.
42. ____________ Only human beings depend on the planets for life.
43. ____________ A galaxy is a system, or group, of many, many stars.
44. ____________ The Sea of Equity is a small sea on Mars.

Answer these questions (each lettered item, 1 point).

45. What are the names of the eight planets? (Don't forget the one on which we live.)

a. ______________________ b. ______________________

c. ______________________ d. ______________________

e. ______________________ f. ______________________

g. ______________________ h. ______________________

Complete this item (each lettered item, 1 point).

46. Name four things not on the moon that are on the earth.

a. ______________________

b. ______________________

c. ______________________

d. ______________________

SCIENCE 409

Unit 9: The Planet Earth

TEACHING NOTES

MATERIALS NEEDED FOR LIFEPAC	
Required	**Suggested**
• a drinking glass • some ice cubes • some flavored beverage • a clear cloth or tissue • a glass • some hot water • 2 ice cubes • a paper towel • a ball (3 to 5 inches) • a light source • a dish (3 to 5 inches)	• 4th Grade Science Experiments video

ADDITIONAL LEARNING ACTIVITIES

Section 1: The Air (Atmosphere)

1. Clouds can be demonstrated by cooling air laden with moisture. Obtain a quart bottle and pour one cup of hot water into it. Put on the bottle cap to keep moisture in the bottle. Then have students observe what happens. A fog should develop. Discuss.
2. Demonstrate buoyancy of salt water vs. fresh water. Use a glass of fresh water and one of salt water. Use various objects (egg, wood, plastic boat, etc.) to illustrate.

Section 2: The Water (Hydrosphere)

1. Secure a large, inflated ball and outline on it the different continents in black water color. Students may be helped to do this activity. Color the water blue. Demonstrate various concepts using the globe. For instance, place a small paper boat on the globe and watch it "disappear" as it moves around the globe.

Section 3: The Land (Lithosphere)

1. Discussion question: Is life on other planets? Have students determine what is required for life on other planets; then see if these requirements are met.
2. Student may get a Christian book on the "Great Flood" and make a report (oral or written).
3. Student may make earth formations (volcanoes, mountains, valleys, plains, etc.) using salt-flour or plaster of Paris.

ANSWER KEYS

SECTION 1

1.1 ionosphere — 50 to 300 miles
1.2 troposphere — 1 to 10 miles
1.3 stratosphere — 10 to 30 miles
1.4 ionosphere
1.5 troposphere
1.6 stratosphere
1.7 Teacher check
1.8 Joseph Priestley
1.9 earth
1.10 mixture
1.11 oxygen
1.12 nitrogen
1.13 one-fifth
1.14 a. red
b. yellow
c. blue
1.15 The glass felt wet.
1.16 Yes, the tissue was wet.
1.17 No colored spot was on the tissue.
1.18 No, I could not taste it.
1.19 They came from the air.
1.20 true
1.21 true
1.22 true
1.23 false
1.24 true

SELF TEST 1

1.01 f
1.02 g
1.03 j
1.04 i
1.05 e
1.06 d
1.07 h
1.08 b
1.09 a
1.010 c
1.011 earth
1.012 protection
1.013 troposphere
1.014 Antoine Lavoisier
1.015 oxygen
1.016 atmosphere
1.017 stratosphere
1.018 ozone
1.019 Any order:
a. air (or atmosphere)
b. water (or hydrosphere)
c. land (lithosphere)
1.020 changing from a liquid to a gas
1.021 water that has changed into a gas
1.022 God created the heaven and the earth
1.023 changing from a vapor or gas into a liquid
1.024 fog forms
1.025 -ated
1.026 -ized
1.027 -ic
1.028 -ing

SECTION 2

2.1 false
2.2 false
2.3 true
2.4 true
2.5 true
2.6 Teacher check
2.7 Teacher check
2.8 Teacher check
2.9 a. lakes
2.10 c. rivers
2.11 b. pond
2.12 a. rivers
2.13 b. glaciers
2.14 "Who covereth the heaven with clouds, who prepareth rain for the earth, who maketh grass to grow upon the mountains."
2.15 God
2.16 rain
2.17 clouds
2.18
1. snow melts
2. lakes fill up
3. evaporation
4. water vapor
5. clouds
6. rain or snow falls

2.19

kindness	payment
readable	coolness
childish	foolish

2.20	Dead Sea	salt in oceans and seawater
2.21	an ocean	Caspian
2.22	3 1/2%	Pacific
2.23	Sea	most salt
2.24	71%	water on earth

2.25 hy' dro sphere, cur' rent, cli'mate, e vap'-o rate', strat' o sphere, mois'ture, at' mo sphere, hor'i zon'tal, pre cip'i ta'tion
2.26 The tiny bits of paper move around.
2.27 The cold water from the ice cube sank to the bottom and warmer water moved up to replace it. As a result, currents were formed that moved the paper.
2.28 a. Matthew Maury
2.29 b. sink
2.30 c. lighter
2.31 b. currents
2.32 "The fowl of the air, and the fish of the sea, and whatsoever passeth through the paths of the seas."

SELF TEST 2

2.01 e
2.02 g
2.03 j
2.04 i
2.05 d
2.06 a
2.07 h
2.08 b
2.09 f
2.010 c
2.011 all of the land or solid part of the earth
2.012 all of the bodies of water on the earth
2.013 all of the air above us
2.014 Matthew Maury
2.015 water cycle
2.016 water vapor
2.017 salt
2.018 cool
2.019 Mediterranean Sea
2.020 Dead Sea
2.021 Any order:
a. Rivers flow in one direction (currents); lakes do not.
b. Rivers have no layers of water; lakes have layers of water.
c. River bottoms are constantly changing; lake bottoms often stay the same.
2.022 Either order:
a. Oceans and seas are both large bodies of salt water.
b. Seas and oceans both have water that contains 3 ½% salt.
2.023 Any order:
a. Seas are smaller than oceans.
b. Many Biblical events are connected with seas, and not oceans.
c. Seas are more numerous than oceans.
2.024 Example: Mr. Maury read Psalms 8:8 and the Spirit led him to think that the "paths of the sea" refers to the currents of the ocean.

SECTION 3

3.1 4
3.2 2
3.3 3
3.4 1
3.5 a. core
3.6 b. mantle
3.7 a. granite and basalt
3.8 c. 4
3.9 Example: This verse means that all things belong to God.
3.10 round
3.11 no
3.12 Teacher check
3.13 a. sphere
b. ground or land
c. feet
d. earth
e. water
f. clouds
3.14 Teacher check

3.15	mountains	flat area less than 2,000 feet above sea level
3.16	volcano	opening in the earth's crust where hot rock comes out
3.17	plain	flat area above 2,000 feet
3.18	plateau	a landform that reaches very high into the air
3.19	glacier	a great body of packed ice

3.20 Examples:
a. rain
b. wind
c. earthquakes
d. volcanoes
3.21 magnetism
3.22 gravity
3.23 air (atmosphere)
3.24 storms
3.25 poles
3.26 sun
3.27 colder
3.28 axis
3.29 earthquake
3.30 moving

SELF TEST 3

3.01 h
3.02 e
3.03 i
3.04 b
3.05 c
3.06 f
3.07 j
3.08 g
3.09 d
3.010 a
3.011 true
3.012 true
3.013 false
3.014 false
3.015 true
3.016 two
3.017 mantle
3.018 Either order:
a. granite
b. basalt
3.019 four
3.020

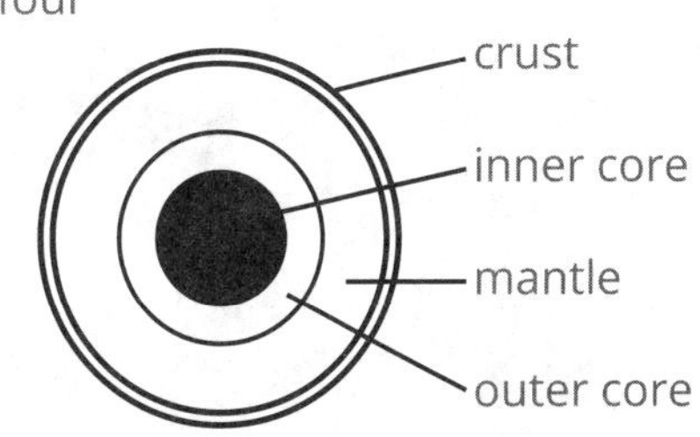

3.021 the water portion of the Earth
3.022 Without oxygen a fire will not burn. Pure oxygen would cause an instant, disastrous fire.
3.023 because of the water cycle
3.024 crust, mantle, and inner and outer cores
3.025 granite and basalt
3.026 Example: The earth moves, slips, or slides along the fault in the crust.
3.027 Example: "The earth is the Lord's and the fulness thereof; the world, and they that dwell therein."This verse tells us that everything is God's.

LIFEPAC TEST

1. g
2. d
3. j
4. b
5. h
6. a
7. i
8. f
9. c
10. e
11. false
12. false
13. true
14. true
15. true
16. false
17. false
18. true
19. false
20. true
21. true
22. false
23. false
24. false
25. false
26. true
27. false
28. false
29. false
30. true
31. b
32. c
33. c
34. a
35. a
36. air, water, and land
37. core (inner core and outer core), mantle, and crust
38. A fault is a break in the crust of the earth where rocks have moved. An earthquake is quick movement of rock along a fault accompanied by shaking.

ALTERNATE LIFEPAC TEST

1. d
2. h
3. c
4. i
5. a
6. g
7. b
8. f
9. j
10. e
11. true
12. false
13. true
14. true
15. false
16. false
17. true
18. false
19. false
20. false
21. true
22. true
23. false
24. false
25. true
26. true
27. false
28. false
29. false
30. true
31. b
32. b
33. a
34. c
35. c
36. Any order:
    a. core
    b. mantle
    c. crust
37. Any order:
    a. air (or atmosphere)
    b. water (or hydrosphere)
    c. land (lithosphere)
38. An earthquake is a quick movement of rock along a fault accompanied by shaking. A fault is a break in the crust of the earth.

SCIENCE 409

ALTERNATE LIFEPAC TEST

NAME ______________________

DATE ______________________

SCORE ______________________

Match each item (each answer, 2 points).

1. _______ plain
2. _______ saturate
3. _______ rotate
4. _______ Lavoisier
5. _______ oxygen
6. _______ atmosphere
7. _______ eclipse
8. _______ plateau
9. _______ precipitation
10. _______ condense

a. an odorless, tasteless, colorless gas
b. complete or partial blocking of light
c. to move around an axis
d. flat stretch of land less than 2,000 feet high
e. change from a vapor or gas to a liquid
f. flat stretch of land more than 2,000 feet high
g. air and other gases around the earth
h. soak thoroughly
i. French chemist
j. depositing of moisture

Write *true* or *false* (each answer, 2 points).

11. __________ Gravity is one of the earth's forces.
12. __________ A particle is the largest part of a substance.
13. __________ Rotation is the movement of the earth around its axis.
14. __________ The troposphere is the lowest region of the atmosphere.
15. __________ A river current constantly changes direction.
16. __________ The earth was created by accident.
17. __________ The hydrosphere includes all bodies of water on earth.
18. __________ Matthew Maury was a famous geologist.

19. ____________ Ponds are usually cooler than lakes.

20. ____________ Cracks where rocks have never moved are called faults.

21. ____________ The earth's core is made up of three kinds of metal: nickel, iron, and silicon.

22. ____________ Glaciers are very powerful to change the land.

23. ____________ Nitrogen has a bad odor.

24. ____________ Revolution is the movement of the earth around the moon.

25. ____________ *Evaporate* means *to change from a liquid to a gas.*

26. ____________ Lava comes from a volcano.

27. ____________ Basalt is a soft rock that chips easily.

28. ____________ All the air above us is called the lithosphere.

29. ____________ The three main parts of the earth are water, air, and volcanoes.

30. ____________ The stratosphere is the region of the atmosphere above the troposphere.

Write the correct letter on the line (each answer, 2 points).

31. The earth's surface is _______ percent water.
a. 82 b. 71 c. 14

32. Evaporation and condensation are a part of the _______ .
a. ocean b. water cycle c. rain

33. Giant ice packs are called _______ .
a. glaciers b. icebergs c. earthquakes

34. A volcano throws out hot _______ .
a. basalt b. granite c. lava

35. The air is made up of _______ nitrogen.
a. 19% b. 50% c. 79%

Answer these questions (each problem, 10 points).

36. What are the three main layers of the earth?

a. ________________________________

b. ________________________________

c. ________________________________

37. What are the three main parts of the earth?

a. ________________________________

b. ________________________________

c. ________________________________

38. What is the difference between an earthquake and a fault?

____

____

____

____

SCIENCE 410

Unit 10: Understanding God's Wonderful Creation

TEACHING NOTES

ADDITIONAL LEARNING ACTIVITIES

MATERIALS NEEDED FOR LIFEPAC	
Required	Suggested
• None	

Section 1: God's Wonderful World

1. Have a group prepare and present a short dramatic skit based on Matthew Maury's discovery of the "paths in the sea" from Psalm 8:8.
2. Prepare and present a similar skit about Benjamin Franklin and his discovery of electricity.
3. Interested students may wish to make drawings of the teeth and jaws of different animals, and show adaptations.
4. Write a short biography of one of the astronauts.
5. Write a short fiction story about a trip to one of the planets.
6. Make a picture showing the "balance of nature." People and animals need oxygen; plants need carbon dioxide. Label the two gases and show how plants help us.

Section 2: God's Changing World

1. Have students investigate the change in the form of a drop of perfume. Place a drop of perfume in a clean bottle and watch it change from a liquid to a gas. Discuss evaporation.
2. Have each member of a group select a different scientist who has been mentioned in the previous LIFEPAC and make up a list of questions about him. Have a quiz based on the questions.
3. Read about Matthew Maury from several sources. Write a short review of his life and share it with the class.
4. Research the subject of oceanography and tell about it in class. Maury was one of the first oceanographers. He discovered one of the "secrets" of the sea from the Bible.
5. Since dust affects rainfall and cloud formation, weathermen measure the dust in the air. You, too, can measure dust. Cover each of two cardboard squares with a six-inch strip of fly paper, sticky side out. Put one square in a box where air currents do not reach it. Place the other square outdoors. After three days, observe and compare the two squares using a microscope.
6. Keep a record of the weather for a month. Test the air for dust every three days. Each time compare the strips. Was there any connection between the amount of dust in the air and the weather?

Section 3: Man's Creative Ability

1. Read about Isaac Newton and tell about what he did.

Section 4: Man's Changing Environment

1. One student may organize a "Clean-Up" committee to clean up the school yard. Also, with permission from the teacher and principal, they may wish to volunteer to clean up a yard for an elderly person or couple after school or on Saturday. An adult should supervise the project.
2. Look up in an encyclopedia, other reference book, or online information about the Venus Flytrap. Tell the class how it traps and "eats" insects. Tell where the plant is found.
3. Search the area around your home or neighborhood for evidence of pollution. Describe what you find and write down some ways to get rid of pollution.
4. Student may catch crickets and keep them in a large glass container for individual or class observation.

ANSWER KEYS

SECTION 1

1.1 false
1.2 true
1.3 false
1.4 true
1.5 true
1.6 false
1.7 true
1.8 false
1.9 true
1.10 false
1.11 a. 1969
1.12 a. Mercury
1.13 c. astronomy
1.14 b. 3,000 to 5,000
1.15 a. telescope
1.16 b. day and night
1.17 b. science
1.18 Any order:
a. sun
b. moon
c. earth
d. planets
1.19 Any order:
a. meteoroids
b. comets
c. asteroids
1.20 10 pounds
1.21 false
1.22 true
1.23 false
1.24 true
1.25 true
1.26 Any order:
a. root(s)
b. stem(s)
c. leaf (leaves)
d. flower(s)
e. fruit(s)
1.27 Either order:
a. to make food
b. to give off oxygen or to provide shelter; enjoyment

1.28
a. carbon dioxide
b. food
c. leaves
d. sun

1.29	God	make food
1.30	decay	created plants
1.31	leaves	take minerals from the soil
1.32	stems	moves up the stem
1.33	roots	carry water to leaves
1.34	water	means to become rotten

(Matches drawn: God – created plants; decay – means to become rotten; leaves – make food; stems – carry water to leaves; roots – take minerals from the soil; water – moves up the stem)

1.35
a. a stem, carries food and water to the plant, holds flowers (blossoms) up
b. receptacle, attaches flower to stem
c. petal, attracts insects and birds
d. pistil, lets flower reproduce
1.36 Teacher check
1.37 h
1.38 f
1.39 i
1.40 j
1.41 b
1.42 e
1.43 d
1.44 g
1.45 a
1.46 c
1.47 false
1.48 false
1.49 true
1.50 true
1.51 false
1.52 true
1.53 true
1.54 true
1.55 true
1.56 false
1.57 true
1.58 false
1.59 true
1.60 true

SELF TEST 1

1.01 l
1.02 f
1.03 e
1.04 h
1.05 m
1.06 k
1.07 a
1.08 b
1.09 j
1.010 c
1.011 g
1.012 Bible
1.013 solar
1.014 soil
1.015 invertebrates
1.016 dance
1.017 Arctic (far North)
1.018 astronomy
1.019 carbon dioxide
1.020 roots
1.021 fruit or apples
1.022 b. decay
1.023 a. oxygen
1.024 c. Mercury
1.025 c. moon
1.026 c. gravity
1.027 b. comets
1.028 true
1.029 true
1.030 false
1.031 true
1.032 false
1.033 Any order:
a. root
b. stem
c. leaf
d. fruit
e. flower
1.034 a. insect
b. mammal
c. insect
d. mammal
e. mammal
f. bird
g. fish
h. bird
i. mammal
j. mammal

SECTION 2

2.1 b. properties
2.2 c. chemistry
2.3 a. chemistry
2.4 a. molecules
2.5 b. a Christian
2.6 b. H_2O
2.7 c. elements
2.8 c. solid
2.9 b. three
2.10 a. elements
2.11 b. different
2.12 c. matter
2.13 a. atoms
2.14 false
2.15 true
2.16 false
2.17 true
2.18 false
2.19 true
2.20 true
2.21 false
2.22 true
2.23 Teacher check
2.24 Teacher check
2.25 Teacher check
2.26 false
2.27 true
2.28 true
2.29 false
2.30 false
2.31 true
2.32 true
2.33 false
2.34 true
2.35 troposphere
2.36 stratosphere
2.37 ozone
2.38 radiation
2.39 altitude
2.40 erosion
2.41 forecasts
2.42 Any order:
a. water
b. plants
c. ice
d. changing temperature
2.43 Any order:
a. thermometer
b. barometer
c. weather vane
d. rain gauge or satellites, anemometers

2.44 f
2.45 g
2.46 j
2.47 d
2.48 i
2.49 h
2.50 b
2.51 a
2.52 c
2.53 e
2.54 Teacher check

SELF TEST 2

2.01 f
2.02 g
2.03 h
2.04 l
2.05 a
2.06 j
2.07 k
2.08 c
2.09 b
2.010 d
2.011 matter
2.012 nine-tenths
2.013 Celsius
2.014 vertebrates
2.015 water
2.016 Bible
2.017 matter
2.018 God
2.019 expands
2.020 sun
2.021 b. elements
2.022 a. Arctic
2.023 c. soluble
2.024 b. barometer
2.025 c. vapor
2.026 c. dance
2.027 true
2.028 false
2.029 false
2.030 true
2.031 true
2.032 false
2.033 true
2.034 a, c, e, Any order:
  a. liquid
  b. Example: water
  c. gas
  d. Example: oxygen
  e. solid
  f. Example: ice
2.035 Examples; any order:
  a. thermometer
  b. barometer
  c. rain gauge or weather vane, anemometer, satellite

SECTION 3

3.1 earth
3.2 bigger
3.3 gravity
3.4 Sir Isaac Newton
3.5 center
3.6 109
3.7 space
3.8 space
3.9 false
3.10 true
3.11 false
3.12 false
3.13 false
3.14 true
3.15 true
3.16 true
3.17 false
3.18 true
3.19 true
3.20 false
3.21 true
3.22 true
3.23 true
3.24 false
3.25 Teacher check
3.26 j
3.27 i
3.28 f
3.29 h
3.30 b
3.31 a
3.32 g
3.33 d
3.34 Conductors are materials that can carry electricity from place to place.
3.35 An insulator is a covering to keep electricity from flowing out of a wire in the wrong place.
3.36 A switch is usually used to break a circuit.
3.37 Teacher check
3.38 Examples; any order:
a. steel pins
b. paper clips
c. needles
d. tin cans
3.39 After the electricity has been generated in the power plant it leaves the generator through wires. The wires connect to other wires that lead to the home.
3.40 true
3.41 false
3.42 false
3.43 false
3.44 true
3.45 true
3.46 true
3.47 b
3.48 c
3.49 d
3.50 e
3.51 a
3.52 Any order:
a. lever
b. inclined plane
c. wedge
d. wheel and axle
e. pulley
f. screw
3.53 wedge
3.54 screw
3.55 lever
3.56 plane
3.57 mesh
3.58 pulley
3.59 Either order:
a. fixed pulley
b. block and tackle
3.60 mechanical advantage
3.61 false
3.62 false
3.63 true
3.64 false
3.65 true
3.66 true
3.67 false
3.68 true
3.69 true
3.70 true
3.71 a-k. Any order:
a. pulley
b. lever
c. wheel and axle
d. hammer
e. wedge
f. oars
g. inclined plane
h. block and tackle
i. seesaw
j. ramp
k. windlass

l-t Any order:
l. crane
m. automobile
n. bicycle
o. ax
p. eggbeater
q. dredge
r. pile driver
s. clock
t. typewriter

SELF TEST 3

3.01 b
3.02 f
3.03 e
3.04 h
3.05 c
3.06 j
3.07 i
3.08 d
3.09 a
3.010 g
3.011 leaf
3.012 gravity
3.013 elements
3.014 Ben Franklin
3.015 sun
3.016 H_2O
3.017 pulley
3.018 Celsius
3.019 invertebrates
3.020 mammal
3.021 c. Newton
3.022 a. troposphere
3.023 a. stand in water
3.024 c. soluble
3.025 b. magnetism
3.026 b. machines
3.027 false
3.028 true
3.029 true
3.030 false
3.031 true
3.032 false
3.033 false
3.034 Any order:
a. lever
b. inclined plane
c. wedge
d. wheel and axle
e. pulley
f. screw
3.035 Example: Astronomy is a true science. It is the scientific study of the universe and galaxies. Astrology is a superstition. People called astrologists pretend to tell a person's future by studying the stars. We are warned against astrology in the Bible.

SECTION 4

4.1 false
4.2 true
4.3 true
4.4 true
4.5 true
4.6 false
4.7 true
4.8 false
4.9 Any order:
a. salt water
b. river
c. field
d. desert
e. forest
f. human
4.10 d
4.11 c
4.12 a
4.13 b
4.14 d
4.15 Any order:
a. land
b. wildlife
c. water
d. oil
e. energy
4.16 pollution
4.17 erosion
4.18 smog
4.19 extinct
4.20 Teacher check

SELF TEST 4

4.01 e
4.02 g
4.03 i
4.04 f
4.05 h
4.06 m
4.07 d
4.08 o
4.09 b
4.010 l
4.011 a
4.012 k
4.013 j
4.014 c
4.015 n
4.016 environment
4.017 biology
4.018 carbon
4.019 Bible
4.020 spider
4.021 electricity
4.022 life
4.023 atoms
4.024 b. grow larger
4.025 c. elements
4.026 a. resources
4.027 a. 100° C
4.028 c. habitat
4.029 b. photosynthesis
4.030 false
4.031 true
4.032 true
4.033 true
4.034 true
4.035 a-e, Any order:
a. lever
b. inclined plane
c. wedge
d. screw
e. pulley or wheel and axle
f-j, Examples:
f. dredge
g. crane
h. pile driver
i. bicycle
j. tractor
4.036 Examples; any order:
a. paper clips
b. needles
c. steel pins
d. iron filings
e. tin cans

4.037 Example: A conductor is a material (such as a copper wire) that can carry electricity from place to place. An insulator is a covering to keep electricity from flowing out of a wire in the wrong place.

LIFEPAC TEST

1. i
2. h
3. k
4. l
5. g
6. b
7. m
8. o
9. f
10. n
11. c
12. a
13. j
14. e
15. d
16. vertebrates
17. extinct
18. gears
19. biology
20. temperature
21. insect
22. plane
23. atoms
24. b. environment
25. a. soluble
26. c. barometer
27. b. root
28. a. an element
29. a. force
30. false
31. true
32. false
33. true
34. true
35. Any order:
    a. root
    b. stem
    c. leaf
    d. fruit
    e. flower
36. Any order:
    a. liquid
    b. gas
    c. solid
37. Either order:
    a. Newton - discovered laws of gravity
    b. Galileo - developed telescope or
    Volta - discovered battery,
    Edison - invented light bulb,
    Oersted - electromagnet,
    Boyle - Christian chemist

ALTERNATE LIFEPAC TEST

1. k
2. n
3. f
4. m
5. a
6. c
7. j
8. b
9. e
10. d
11. g
12. h
13. i
14. o
15. l
16. water
17. superstition (false science)
18. troposphere (atmosphere)
19. pollution
20. rays
21. atoms
22. gravity
23. soluble
24. c. gained
25. b. element
26. a. environment
27. c. solution
28. a. extinct
29. c. sun
30. false
31. true
32. true
33. true
34. false
35. Examples, either order: Edison, Newton
36. Either order:
    a. liquid
    b. gas
37. Any order:
    a. roots
    b. stems
    c. leaves
    d. flowers or blossoms

SCIENCE 410

ALTERNATE LIFEPAC TEST

NAME ______________________

DATE ______________________

SCORE ______________________

Match each item (each answer, 2 points).

1. _______ temperature change
2. _______ conductor
3. _______ circuit breaker
4. _______ community
5. _______ mosquito
6. _______ wedge
7. _______ Volta
8. _______ chemistry
9. _______ gears
10. _______ erosion
11. _______ Edison
12. _______ ferns
13. _______ gravity
14. _______ thermometer
15. _______ barometer

a. insect
b. study of matter
c. inclined plane
d. wearing away by wind and rain
e. wheels with teeth that turn
f. switch
g. electric light
h. forest community
i. holds things on earth
j. battery
k. one cause of weather
l. measures air pressure
m. habitat for living things
n. copper wire
o. measures air temperature

Complete these statements (each answer, 4 points).

16. H_2O is a symbol for ________________ .

17. Astrology is a ____________________________ .

18. The air nearest the earth is called the ________________________________ .

19. Waste and dirt in the air and water is called ______________________ .

20. A layer of ozone above the earth protects us from ultraviolet ______________ .

21. Molecules are made up of very small particles called ______________ .

22. Sir Isaac Newton discovered the laws of ________________ .

23. Because sugar dissolves in water, it is said to be ___________________ .

Write the correct letter and answer on the line (each answer, 3 points).

24. Mechanical advantage is the force ______________ by using machines.
a. lost b. wasted c. gained

25. Gold is ______________ .
a. very light b. an element c. soluble

26. All conditions and influences that have to do with the growth of things make up the ________________________ .
a. environment b. circuit c. invertebrate

27. When sugar is dissolved in water it is called a ______________________ .
a. symbol b. potion c. solution

28. Wildlife that no longer exists is said to be ___________________ .
a. extinct b. happy c. famished

29. The center of the solar system is the ______________ .
a. earth b. moon c. sun

Write *true* or *false* (each answer, 1 point).

30. ____________ Ecology is a branch of the science of chemistry.

31. ____________ A cricket belongs to the insect family.

32. ____________ A bulldozer is a complex machine.

33. ____________ The turnip plant stores food in its root.

34. ____________ Molecules are made up of large particles called sepals.

Answer these questions (each numbered item, 5 points).

35. What are the names of two important scientists?

a. ____________________ b. ____________________

36. What are two of the three forms of matter?

a. ____________________ b. ____________________

37. What are the four main parts of a plant?

a. ____________________

b. ____________________

c. ____________________

d. ____________________